TRACING THE ADVANCE OF TECHNOLOGY AND DELVING INTO TECHNICAL THINGS

KERWIN MATHEW

TRACING THE ADVANCE OF TECHNOLOGY
AND DELVING INTO TECHNICAL THINGS

PREFACE

It is difficult to keep up with technology. It is even more difficult to be able to keep up with technology and at the same time be capable of managing technology well.

This book attempts to provide the technical executive, e.g., the R & D engineer or production engineer, who has to manage technology, with some knowledge which he could find useful and implement in his organization.

Kerwin Mathew, Ph.D., PE, CMfgT, CPM

CONTENTS

1 PORTABLE CAMERA THAT PRINTS

A Japanese stationery goods manufacturer, King Jim, has developed the Da Vinci, the world's first digital printing camera/portable copying machine which is capable of reproducing three dimensional objects and printing as many copies as required.

The equipment looks like the remote control unit for a television set and is easy to operate. To produce a photograph one simply presses the yellow button followed by the PRINT button. The photograph is monochrome and can be changed into any one of 60 printed forms.

It acts as a filing aid in offices.

2 THE PALM TOP

Sony Corporation has developed a new, keyboardless computer known as Palm Top PCT - 500. The computer is about the size of a book, has the same screen resolution as the MacIntosh, weighs 1.3 kg., comes with a complete set of software and is equipped with a special pen.

The computer opens and lays flat. The user writes on the screen. The machine translates his handwriting using "fuzzy logic". It can handle 3,500 Japanese characters and simple English. More languages are expected to be added to the list.

3 GLOBAL ENERGY NETWORK EQUIPPED WITH SOLAR CELLS AND INTERNATIONAL SUPERCONDUCTOR GRIDS (GENESIS)

The world is mainly dependant on fossil fuels such as coal, oil and natural gas for its energy supply. However, such fuels are limited in supply and cause environmental pollution. But the sun's energy, solar energy, is limitless in supply and does not disrupt the environment.

Sanyo Electric has pioneered research in and commercialization of solar cell technology. It has begun the production of amorphous silicon solar cells in 1980 and is a world leader in the field.

It has come up with an ambitious plan to build large solar power plants in deserts, on plains, or even on suitable locations on the sea - each of these facilities would be connected by a global transmission grid made of superconducting electrical cables, known as Project GENESIS (Global Energy Network Equipped with Solar Cells and International Superconductor Grids). This would allow power to be supplied throughout the world at any time, regardless of weather conditions. The surplus energy could be stored in superconductor storage facilities.

The potential of solar energy is very great. As it is, with the present technology, only about ten percent of the sun's energy falling onto a solar cell's surface is converted to electricity. If a way could be found to utilize most of the remaining 90% the world would be well on its way to solving its current energy problems.

4 HIGH TECH DEVELOPMENTS WILL AFFECT THE MEDIA INDUSTRY

Videocassette recorders with scanning sensors that delete commercials from video-taped programs have already been introduced in Japan.

Passive "people meters" which could electronically identify the facial features of television viewers, making it easier to monitor their viewing habits, will be available.

There will be personalized magazines which are delivered by computer. Newspapers will be transmitted by facsimile.

A study conducted by Backer Spielvogel Bates predicts that technological developments in the next ten years such as these will drastically change the way advertising is created and the way media are used.

For one thing many of these changes will help advertisers to reduce media costs or allow marketing managers to target certain consumers more precisely.

Computer-delivered magazines will allow publishers to customize editions for various segments of the population. This means more work for media agencies as they will have to create different media versions to appeal to each group.

Facsimile transmission of newspapers will enable advertisers to reach their audience more easily. Business-to-business advertising in trade publications is likely to be supplanted by newspaper advertising.

It will be possible for advertisers to find out through "people meters" who is watching what commercials second-by-second.

Videocassette recorders with commercial sensors will probably coerce media agencies to create special commercials that will have no sound and will appear in a corner of the television screen during a program.

Advertisers are likely to get what they pay for.

5 LATEST IN SOFTWARE

Drexler Technology Corp. of America has developed a new data storage medium, a card, which can hold about 1,200 pages of text. The card has a mirror-like amorphous crystal surface. Some Japanese companies are trying to commercialize the technology.

6 HIGH TECH WAREHOUSING SYSTEM

Aiwa S'pore, a Sony subsidiary which has all along been only making audio products here, has started producing video equipment as well.

It has invested about $130 million in three factories here which manufacture audio equipment such as head-phones and mini hi-fi sets, accounting for 40 percent of its worldwide production.

The company has a Central Distribution Center at the YCH Distripark in Tuas. The Center has a computerized Warehouse Data Communication System (WDCS) which uses mobile terminals linked by UHF radio to a computerized management system. The WDCS relies on hand-held laser guns and reduces human errors in storing and retrieving goods. It allows on-line stock control with real-time inventory reporting. As compared to the more conventional methods it gives about 40 per cent savings in operating man hours.

7 APPLE COMPUTER'S WIRELESS PC COMMUNICATIONS USING RADIO FREQUENCIES

Apple Computer Inc. has approached the Federal Communications Commission (FCC) to put aside radio frequencies to allow its personal computers to receive and transmit data without wires.

Computer communication by radio gives more flexibility. The company thinks this mode of communication will be a major trend in personal computing. It is requesting frequencies that could be used by all the other computer companies as well, so that wireless local-area networks (LANS), which are systems that link computers within an office or building, are possible.

LANS generally use coaxial cables or telephone wires which may require expensive and time-consuming rewiring , especially in older buildings. Apple wants to change all this and is asking for frequencies in the region of 40 megahertz that would allow the building of lower-cost networks and the interference-free transmission of computer data.

The FCC would take at least one year to decide on the request.

If Apple's request were entirely approved by the FCC it would be possible to have several local area networks in a given area with transmission speeds of about 10 million bits per second and transmission distances of about 45 meters.

8 THE FABULOUS SUPERCOMPUTERS

Supercomputers, which have been developed mainly for military objectives, can perform billions of mathematical operations in a second, store and retrieve large amounts of data and perform complex simulations with great accuracy. They are thousands of times more powerful than desktop computers and cost US$3 million to US$30 million each.

Once used almost exclusively to construct nuclear weapons and crack military codes, they are finding a number of new applications as environmental tools as was evident in the war fought in the Persian Gulf, where they were employed to evaluate the damage to the environment from the burning of vast quantities of oil; the supercomputer simulates oilfield fires showing their effect on the environment and weather, helping greatly in efforts to assess the real damage.

Carnegie-Mellon University scientists have demonstrated that smog stays put high above the ground, moving lower each morning as warming air circulates, adding to the new day's pollution, by using a supercomputer. This knowledge resulted a broadening of anti-pollution policies aimed principally at car emissions, and led Los Angeles to adopt the strictest pollution controls in the country.

Policy-makers can use supercomputers to experiment quickly by creating a series of hypothetical situations and measuring the effects of variable factors.

Supercomputers are widely used for numerous engineering and scientific tasks such as the simulation of automobile accidents to learn how to design safer cars, the creation of aeroplane models to improve their aerodynamic efficiency and the simulation of the behavior of molecules to create more effective medications.

Cray Research Inc., the number one supercomputer maker in the US, has taken on large and increasingly more environmental projects.

Its chairman and chief executive has dubbed this "the greening of the supercomputer".

9 JAPANESE SOFTWARE SUPERIORITY VIS A VIS AMERICAN SOFTWARE

The Americans have watched the Japanese taken the lead over them in key manufacturing industries and it has dawned on them that the Japanese are going to outdo them in software writing as well.

A comprehensive study of the Japanese software industry has found that Japan's best companies are more productive and write software with much fewer bugs than their US counterparts.

The severe shortage of programmers and fussy domestic customers demanding customized software have ensured that Japanese software writing reached a high level of skill.

Companies such as Hitachi Ltd., Fujitsu Ltd., NEC Corp. and Toshiba Corp. have established factories in which thousands of programmers toil together, making use of standardized training, design tools and methods to avoid duplication of work and to reduce mistakes. The best US companies also employ similar techniques.

However the Japanese are more superior than the Americans in their software technology. It has been reported that Japan's best companies are 30 to 40 percent more productive per worker and write software which has 25 to 50 percent fewer bugs than their US counterparts, though their software engineers are less creative than the latter's.

10 THE PHONE BILLING MONITOR

Four students of a tertiary institution here have invented an automatic phone billing monitor. Later in the year the Telecom Authority would bill phone users by the actual call-time. This invention has not come at a better time.

The small phone billing device monitors the duration and cost of all outgoing local and IDD phone calls, and gives the total cost for the month or other periods, displaying such information on a small screen.

It also keeps track of and records the number of incoming calls and automatically redials these numbers for the owner later with the touch of a button.

It can store up to 10 telephone numbers and display the current IDD charges for 20 countries, the time and the date; the potential storage capacity is much greater.

It is portable, can easily be plugged into the phone and is powered by a plug in the wall.

It is reported that several manufacturers have shown interest in the $150 prototype. It is believed that with mass production consumers could obtain it at a price of less than $100.

11 SOME INVENTIONS BY STUDENTS WITH COMMERCIAL VALUE

1. Paper made from water hyacinth. The water hyacinth is blended. The pulp is extracted with chemicals and a filter. It is then turned into rough paper.
2. Cement made from blended rice husks. Rice husks are mixed with ordinary portland cement to produce a more durable formula.
3. Artificial hand and finger joints. These are made from a pliable and durable plastic and are for the disabled and the arthritic sufferers.
4. Semi-automatic duck gizzard cleaner. This can replace the current manual method.
5. Fiber-enriched noodles. This food reduces the risk of chronic diseases such as colon cancer and coronary heart disorders.
6. Solar-powered catamaran. This has a solar panel which is mounted at the top and which powers two motors and charges two batteries in the hull.
7. Low-cost automatic page-flipping machine. This is operated by a foot pedal and is for the disabled.
8. Portable vehicle detection system. This monitors illegal parking and takes pictures of cars.
9. Karate punch-and-kick recording machine. This is for martial arts and gymnastic training.
10. "Stand-up" wheelchair. This enables the disabled to have eye-to-eye conversations.

12 HI-TECH SPORTS

Bridgestone Hi-Tech Sports Centre here has brought in the Science-Eye, an $100,000 equipment which will help golfing enthusiasts, both novices and experts, to analyze their deficiencies.

It was originally developed to test the efficiency of golf balls and is a combination of electronics, computer, television and video technology.

The test area, which measures 74 sq. m., has two cubicles net are enclosed by netting and a target, and green carpeting that simulates an actual golf course. The first cubicle is for warming up while the second one is for teeing off.

There are five closed-circuit cameras to capture the golfer's posture. The impact of the swing is registered by the computer. The Science-Eye will analyze the hit on screen, provide still pictures of the golfer's stance and produce a print-out of the club's head speed, ball velocity, hitting accuracy and launching angles.

The analysis will allow Hi-Tech Sports Centre to recommend the clubs most suited to the golfer.

However, the Centre will not advise golfers on how to improve their golf. The latter will do well to get golfing advice from the professionals.

13 NEW PC AND WORKSTATION FAX SERVER FROM XEROX

Xerox Corp. has introduced an integrated fax server, the Xerox LAN/Fax Express 21, which allows Novell Inc. local area network (LAN) users to send and receive facsimile transmissions directly at personal computers or workstations.

The machine consists of a Xerox Telecopier 7021 plain-paper fax terminal and the LAN/Fax Express hardware accessory and software.

14 COMPUTER-AIDED MEDICAL ILLUSTRATION

Powerful computers are being used to create detailed three-dimensional pictures of the body on computer screens.

It is reported that the whole process is like working with the real object. It is possible to go in and explore within the three-dimensional space and turn it around and look at it.

Medical illustrators used three-dimensional computer graphics taken from the entertainment industry.

Take the case of a growing boy who lost his nose to cancer and for whom a new nose is to be constructed. The "nose designer" creates three-dimensional images by scanning objects with a laser device to read the coordinates and entering the data into computers.

It is reported that the work being done with three-dimensional imaging is changing how medical illustrators learn and work and that three-dimensional imaging is good for teaching.

It is hoped that three-dimensional imaging will supplement books and cadavers and, at some point in time, replace illustrations in books.

15 THE ELECTRASCAN

The Electrascan is used to assess the level of toxic heavy metals and is the first in the world.

It does not require much technical training to use the monitor to trace the concentration levels of metals in remote locations, a process which previously required samples to be sent to a laboratory for analysis. Besides being simple, it reduces analysis costs.

The bulkier prototype model was developed by Eutech Cybernetics, a company which specializes in computer-based process control and monitoring instrumentation started by a former National University of S'pore lecturer with equity participation by Novo Technology, a subsidiary of S'pore Institute of Standards and Industrial Research (SISIR). SISIR scientists later miniaturized the circuit board, giving it an impressive and compact design and reducing it to about the size of the palm.

For this pioneering product Eutech has so far secured more than $3 million worth of orders in the USA.

16 INTERFIRM COMPARISON (IFC)

An Interfirm Comparison (IFC) seminar has been held at the National Productivity Board (NPB).

Local companies have been urged to participate in the IFC concept, a new management technique here that encourages companies to share information to identify common areas for productivity growth.

The findings of three IFC studies which have been carried out by the NPB and the NTI-Peat Marwick Entrepreneural Development Centre (ENDEC) in the previous year have been released at the seminar.

In one of the studies, industry-wide statistics on wages for the garment-manufacturing sector were compiled. The information were taken from 35 companies and included the average hourly earnings for various categories of workers in 1989 which served as a useful indicator for the industry.

Useful information on productivity related to inventory, labor and selling space were obtained from the other two studies involving six local department stores and seven apparel retailing outlets.

A comparison of the above data with statistics from the United States showed the following:

(1) Local department stores performed twice as well as the Americans in manpower and space
    productivity.
(2) The inventory productivity of local retailers lagged behind that of the Americans.

The following conclusion was drawn from the comparison: The companies here had to improve on their warehousing, delivery systems and distribution arrangements.

17 NATIONAL IT MASTERPLAN

A steering committee comprising of 11 sectors of the economy, viz., the government and ten industries - manufacturing; financial; construction and real estate; retail, wholesale and distribution; leisure and tourism; transportation; healthcare; education; publishing and media and IT - headed by the chairman of the National Computer Board, has drawn up the National IT Masterplan with the object of exploiting IT for an economically vibrant country. Representatives from the Economic Development Board, the Institute of Systems Science, the Institute of Policy Studies, the National Science and Technology Board and the Telecommunications Authority are also in the committee.

It is thought that IT will bring about a better quality of life and that business success will be more and more dependent on how information within a company, the country and across the world is processed and managed.

International experts will be roped in to help develop relevant areas in the Masterplan. To fine-tune the needs of each sector there will be sub-committees, which will meet about five to six times in the year to discuss the nitty-gritty, while the steering committee will play a coordinating role.

The Masterplan will be reviewed yearly to find out the areas where changes and modifications are needed.

It should ultimately propel the country into the mainstream of high-level technology.

18 THE COMPUTERIZED PEN

An equipment of powerful simplicity, the computerized pen, has made its appearance. It has a large potential market. Hardware and software producers have gone full swing into developing electronic note pads that use a pen instead of a keyboard to enter and call up information.

Millions of consumers who do not have the time to digest or are in awe of computer technology could be attracted by the simplicity of the pen and take the plunge into the complex world of computers.

Go Corp. of Foster City, California, has announced that it would begin sending software developers its pen-point operating system for "mobile" computers which will be operated by pens instead of a keyboard or a mouse.

It has asked a large number of companies, e.g., Borland International Corp., Lotus Development Corp. and Word-Perfect Corp., to write application programs, software that performs specific tasks, for its pen-point operating system.

It has the backing of computer hardware giants, Apple, NCR Corp. and IBM.

Besides Go there are several others who are developing or have developed pen-based systems. Microsoft Corp. is developing its own pen-based system to be used together with Windows, the popular program that makes the DOS operating system much easier to use. Tandy Corp. has a notepad computer, the grid pad, which looks like an "etch-a-sketch" with a pen and cord attached. Wang Laboratories Inc. has a pen-based system known as Free-style, which hooks up to a PC. However both the Tandy and Wang systems are limited in the range of what they can do.

The realization that there is a huge number of non-computer users has spurred the development of the pen-based system.

According to the forecast of Infocorp., a market research firm, about 50,000 notepad computers would be sold in the US in the year and that number is expected to increase to more than two million in the next few years - by then the number of notepad computers used would provide a great demand for application programs.

As the pen-based computer is more portable, since the cumbersome keyboard is no more, many field workers would each find it practical to tag along one while out in the field.

A market research has indicated that there is a bright future in the US for the system - at least 25 million non-computer users might use it.

19 THE UNICODE

A consortium has been developing and promoting a new code, known as the Unicode, which is to be the lingua franca for the electronics age. Its 12 members include many top computer companies such as Apple Computer, Microsoft, IBM, Xerox and Sun Microsystems.

The Unicode would make it easier for people in different countries to contact each other through electronic mail; there would not be any language barrier. It would also make it easier for software companies to come up with programs that can work in different languages.

Information is represented in computers as a series of ones and zeros or as digital bits. The ASCII (American Standard Code for Information Interchange) is the most widespread system. It represents each letter and symbol as a sequence of eight zeros and ones, e.g., 10111001 represents the letter Y.

As ASCII cannot handle special characters used in other languages some countries have to design their own codes. For example, Europe has its own 8-bit code while Asian countries such as China and Japan have their own codes to represent the thousands of different characters in their languages. As there are only 256 different 8-bit sequences of zeros and ones ASCII cannot be used to represent characters in all these languages.

However, the Unicode would represent letters and symbols by a sequence of 16 zeros and ones, instead of eight, allowing for 65,536 different combinations. This would make it easier, for example, to develop a word-processing program which works in many languages.

The Unicode, which now includes sequences for 27,000 characters, is developed because the consortium thinks that its overseas markets are becoming more important. It has been under development since 1989.

20 INMARSAT-C

S'pore Telecom has launched the Inmarsat-C, which is the world's smallest portable satellite system, consisting of a cone-shaped antenna and a light-weight electronics unit that together weigh just 12 kilograms.

S'pore is the first ASEAN country to commercialize the service. The service will be mainly used by ship-owners. Transport and shipping staff, journalists and travelling businessmen can now send telex messages and data by connecting their personal computers or lap-tops to the portable terminals which cost between US$8,000 and US$10,000 each. The service is linked through a global satellite system. A more expensive version is the Inmarsat-A system which is used on board large sea-going vessels such as liners, tankers and bulk carriers.

The traffic charge for the Inmarsat-C service is $1.90 per kilobit or $6.10 per telex minute for sending messages from the mobile satellite terminal to shore and vice versa. The charge is double for satellite terminal to satellite terminal communications.

The services that are presently available include message status checking facility, telex and distress alert.

21 S'PORE INSTITUTE OF STANDARDS AND INDUSTRIAL RESEARCH (SISIR) HAS BECOME A ONE-STOP TECHNOLOGY CENTER

Besides setting standards SISIR has been active in research and development. It is aiming at becoming a technology supermarket which provides premium items in areas of standards, certification, research and development. It has gone into "hard technology" areas such as materials technology and product and process technology.

In 1986 it spent $6.5 million on research and development. Today, the figure stands at $16.7 million. The bill for statutory activities such as standardization, testing and certification has gone up from $7.8 million in 1986 to $10 million today.

In 1988 it opened a marketing and business division to enhance its image.

Its Technology Diagnostic Centre (TDC) in the SISIR Building is a walk-in diagnostic clinic for small and medium-sized companies. Its $1-million Information Technology Evaluation Centre (ITEC), which is a joint venture with British software house, Admiral Computing Group, tests and evaluates software quality. Its incubator scheme provides start-up companies and innovators with the infrastructural support and technology which are normally available only in big companies. It has a technology transfer division which specialises in sourcing out new technologies from around the world to check their marketability, thereby helping the institute in charting its own research and development plans.

Beside local companies, multi-national corporations and government bodies, its clientele also includes overseas organizations.

It is establishing an international network which will give it more credibility, while at the same time it is helping the local businesses. This network includes the following:

(1) Accepting the ISO 9000, which is a quality management standard recognized internationally, as the benchmark for its Good Manufacturing Practice Scheme.
(2) Acting as a testing agent for as many as five standard institutes from countries such as Japan, Britain and Canada.
(3) Joint ventures, e.g., with AT & T to provide quality management consultancy and with S'pore Food Industries to look into food packaging technology.
(4) Helping international purchasing offices (IPOs) in S'pore to test products before orders are made.

The institute has recently become self-financing, i.e., it only receives government grants for statutory projects like testing and certification, while all its other projects are funded by clients.

Its immediate task is to be more active in the international standards program and to find out what is the latest in these areas so that local products will find a way to meet these standards and sell in the world market.

It does not rest on its laurels. It is looking ahead and predicting the technological needs of the future, and has gone into such new areas as electromagnetic compatibility assessment, environmental technology and food bio-technology.

22 ELECTRICAL AND ELECTRONIC COMPONENTS AND PRODUCTS

This field seems to have the most innovations. The engineer or purchaser is likely to have difficulty keeping up with the latest products and technology. Products are improving all the time. For example, micro-processors are becoming better and faster, relatively new technology such as fiber optics and lasers are finding wider applications, programmable logic controllers (PLCs) are playing a greater role in instrumentation and printed circuit boards (PCBs) are getting more into surface mount technology whereby more functions are "compressed" into a single board; there is even research and development work on organic circuits, which will be a far-reaching area of electronic technology. There is much research and development work on power switch modes, uninterruptable power supplies, disk drive technology, integrated circuits, communication systems and others. USA and Japan are leaders in the field, with countries like Korea and Taiwan following behind.

23 AUTOMATION SYSTEMS

Servo driven actuators, robots, computer-integrated systems and equipment with AI (artificial intelligence) all fall within the scope of this product category. Automation systems have a very wide application. They are in use in the electrical and electronics; precision engineering and metal working; shipbuilding and ship-repair; tool, die and mold making; food and beverage; chemicals and chemical products; wood and wood products; textile, apparel and leather products; paper and paper products; printing; plastics; and warehousing and materials handling sectors in the industries. Today, terms like CIM (Computer-integrated Manufacturing) and FMS (Flexible Manufacturing System) are familiar sounding as such automation systems find a wide usage. Both the US and Japan lead the field and are about equally strong, however, Japan appears slightly ahead of the US in robotics; countries like Germany, Australia and Sweden also supply robots and automation systems. Automation cuts costs and reduces labor requirement. A sophisticated automation system can result in unmanned operations which can well be the trend in the 21st. century.

24 QUALITY TESTING

Metrology and test instrumentation come under this category; measuring instruments such as micrometers, push-pull gages, profile projectors, multi-meters, oscilloscopes and many others are covered. Quality standards as exemplified by those of UL and CSA are included. Quality control techniques are also included. For example, TQC (Total Quality Control) is practiced in many companies, especially Japanese ones, which are noted for their TQC techniques; its popularity appears to be on the increase. There has been hype about zero defect quality as well.

25 MATERIALS ENGINEERING

This sector covers purchasing, inventory control, production planning and control, materials handling and warehousing. The materials function is a specialized and important one. Material Requirements Planning (MRP) and Manufacturing Resource Planning (MRP II), which rely on high capacity computers, are two production planning techniques now widely used. JIT (Just-In-Time) is a production control technique that is becoming popular despite some misgivings. All these three techniques have been popularized by the US; in fact, MRP and JIT are nothing new, these methods have been used before, the Americans just adopted them and attached the above-mentioned names to them. Sophisticated warehousing systems, including automated ones, are found in many companies.

26 NOVELL'S OPEN DATA-LINK INTERFACE (ODI)

ODI represents Novell's newest interface for a workstation and a network. A distinct advantage of ODI is that it could support multiple protocols and multiple LAN adaptors concurrently. It would therefore be possible for two protocols such as TCP/IP and IPX to be used on the same workstation sharing the same board. The process of linking the workstation to the LAN driver would be simple if a LAN adaptor (network board) meets ODI specifications. The NET.CFG file would replace the traditional SHELL.CFG file (as used in login). The contents of a typical NET.CFG file would comprise statements previously used in the SHELL file and input parameters for the following programs which are called by the AUTOEXEC file:-

(i)   IPXODI   The IPX protocol stack
(ii) TCPIP   The TCP/IP protocol stack
(iii) LSL   The Link Support Layer
(iv) NE 2000   Multiple Link Interface Driver
(v)   NETX   The NETX shell

Four sets or layers of files provide the information required by the workstation to establish network communications, in the AUTOEXEC file. The first layer in NET.CFG which consists of the Link Support Layer (LSL) program enables the workstation to communicate over several protocols. This directs the traffic between the LAN driver and the protocol stacks, acting as a traffic cop. The second layer which is the Multiple Link Interface Driver is an adapter board LAN driver that accepts information packets from multiple protocols. The third layer comprises the protocol stack files such as TCPIP and IPXODI. They manage communications among other network nodes. The fourth layer is the NetWare shell program, e.g., NETX.

ODI is also able to dynamically configure the work-station interface. It is possible to configure as the need arises without having to regenerate the workstation files for this or other workstations. The variety of workstations and protocols that could now connect to the network is almost limitless.

27 USES OF PERSONAL NETWARE

Changes In Networking
Novell, the developer of NetWare, has developed the Universal NetWare Client software in Personal NetWare. One could have access to other PCs in one's workgroup and to the central file server, all at once, with this software installed on one's computer. Peer-to-peer features of the Universal NetWare Client software allow one to have access to working files for a project that might be on a co-worker's PC, while other resources such as gateways to host computers and regular file backups could be handled through the central file server. This arrangement allows you to use a file or printer regardless of where it is.

The Universal NetWare Client Connects You Better Than Ever Before
Unlike in the past, when each type of network has its own set of files for the client PCs, the Universal NetWare Client today gathers all the different support files and connection methods into a single, easy-to-use package, a kind of "one-stop shopping" for all one's network resources.

It is now common to have many servers supporting scores and hundreds of users in one place. The network administrator has to carefully keep track of each user and his or her needs on each server, maintaining separately the access profiles and login scripts, one set per server.

The client software could gain access to multiple server resources automatically with Personal NetWare and the Universal NetWare Client. The same resource list would be available every time the user logs in after a profile and resource list is set up in Personal NetWare.

28 THE UNIX OPERATING SYSTEM

UNIX is probably the most well-known multi-tasking operating system within the micro-processor and mini-computer environment. It is in large measure a result of the work done by a consortium comprising of General Electric, Bell Telephones and Massachusetts Institute of Technology, formed in the late 1960s, to develop MULTICS (MULTIplexed Information and Computing Service). MULTICS was aimed at generating software which would give many users simultaneous access to the computer.

After the development of the C programming language, UNIX was rewritten in C in 1972, which was a major departure for an operating system. The advantage of UNIX is that it protects the user from the computer system it is running on, from having to know the exact location of the memory in the system, what a disk drive is called and other similar information. Many parts of UNIX are of a logical nature - they could be seen and used by the user, though their actual location, structure and functionality are hidden. For instance, if a user wants to run a 20 Mbyte program on a system, UNIX uses its virtual memory capability to get the machine to behave logically, just like one with enough memory, although the system might have only 4 Mbytes of RAM installed. It is possible for the user to access data files without knowing whether they are stored on a floppy or a hard disk, on another machine many miles away, or connected through a network. Using its facilities, UNIX presents a logical picture to the user, at the same time keeping the more physical aspects from view.

The filing system of UNIX is hierarchical in nature. It has all the data files, programs, commands and special files which allow access to the physical computer system. The files are usually divided into directories and subdirectories. The file system commences with a root directory and divides it into subdirectories. There could be subdirectories which continue the file system into further levels and files that contain data, at each level. A directory could have both directories and files. The file system stops at that level for that path when there are no directories.

A file's location in the hierarchy is described by a file name according to the path taken to locate it, commencing at the top and working down. This is often referred to as a tree structure. (Turned upside down, the structure looks like a tree, starting at a single root directory - the trunk - and branching out.)

The UNIX system revolves around its file structure. All physical resources are accessed as files. Commands also exist as files. The physical file system comprises the mass storage devices such as floppy and hard disks, which are allocated to parts of the logical file system.

29 THE OS/2 SOFTWARE

OS/2 was created by IBM and represents a clean break from DOS, the Microsoft software; OS/2 has always been a much better software than Windows. It is a pre-emptive, multi-tasking operating system and has a 32-bit design (whereas Windows is a 16-bit software which is not pre-emptive and which simulates multi-tasking). As compared to Windows, it is more robust and more powerful when run on the correct computer. However, it suffers the disadvantage of having a lack of backward compatibility. However, Version 2.1 of OS/2 has to a great extent diminish this compatibility problem; it could run both Windows and MS-DOS programs. Despite this, it is still not popular with computer-users. IBM has spent plenty of time and money incorporating multimedia capability into Version 2.1 of OS/2. In reality, this 32-bit multi-tasking program is much suitable for multimedia development.

Multimedia puts much strain on an operating system and often requires a program to perform two tasks simultaneously. OS/2 Version 2.1, because it could multi-task and is particularly good at handling video images, is eminently suitable for applications where a computer has to play back simultaneously a mixture of audio, video or graphics, with everything sequenced correctly and without any quality loss.

OS/2 has built into it a comprehensive set of multimedia tools known as Multimedia Presentation Manager /2 (MMPM). Media players which allow musical digital instrument interface (MIDI) files to be played, a data conversion package which allows images and audio files to be converted and a digital audio player/recorder are included in this toolset. There is also a Media Control Interface (MCI) panel like Windows in OS/2 Version 2.1. This makes it easy for software developers to write their multimedia software to work with OS/2. This panel also makes it possible for the correct media player to be loaded when a multimedia file is loaded. OS/2 Version 2.1 uses the MCI panel to integrate device drivers for multimedia hardware and software into an operating system.

30 LOTUS SMARTSUITE

Lotus SmartSuite represents a complete set of office software in one box. With Lotus SmartSuite, it is unnecessary to have any other software, except DOS and Windows, in order to handle all of one's computing needs, such as Ami Pro word processor, 1-2-3 spreadsheet, Approach database, Freelance presentation graphics, Organizer personal information manager and a few smaller applications.

Lotus SmartSuite is indeed designed to work well together as all the software comes from one company. It provides a consistent user interface. Therefore, the various applications have the same look and "feel". Moreover, the applications also share data well. It is easy to import names and addresses from an organizer into a database, turn an outline developed in the word processor into a nearly instantaneous presentation and include a range or chart from a spreadsheet in a word processing document.

Many integrated office software, such as Framework, Symphony and Microsoft Works have lost their popularity as the included applications did not provide as many features as stand-alone applications. For instance, the word processing function of an integrated program might offer only 50 to 75 percent of the features found in a major stand-alone application. Lotus SmartSuite, on the other hand, comprises of stand-alone applications that have been developed separately, in a highly competitive marketplace. Its stand-alone applications are top-ranking, some of the best available. More features have been added to make them work well together when they were packaged together.

It is the aim of Lotus to have their office applications perform so well together that multiple products could easily be called on to solve a business problem, hence, Working Together - a Lotus goal. One of the objects of Working Together is to make sure that all SmartSuite products have the same look and "feel" - if one knows how to use one, one would feel comfortable with the others. There is tight integration among the products. This allows each to take advantage of the most powerful features of its sister products. For instance, Approach is a much better software than 1-2-3, but many prefer the latter for use in developing tables of database records as they are more familiar with it. However, the tight integration of Working Together enables 1-2-3 to call on Approach's superior facilities for such tasks as generating dynamic crosstabs and printing mailing labels. The Approach dialog boxes could be made to appear right in the 1-2-3 window as though they were a part of 1-2-3.

31 SIMPLE GUIDE TO C++ PROGRAMMING

Getting Started

1)   C++ is an efficient, compact, fast and portable programming language, its object-oriented heritage bringing a fresh programming methodology designed to cope with the escalating complexity of modern programming tasks.

2)   C++ brings together two separate programming traditions - the procedural language tradition, represented by C, and the object-oriented language tradition that is represented by the enhancements C++ adds to C.

3)   C++ implements objects and classes, and new features, e.g., multiple inheritance, can be found in Version 2.0 of the language.

4)   It represents a new programming paradigm, object-oriented programming (OOP).

5)   Object-Oriented Programming

5.1) In C++, a "class" specification describes a new data form which corresponds to the essential features of a problem and an "object" is a particular structure constructed according to that plan.

5.2) The OOP approach to program design is to first design classes which accurately represent those things with which the program deals.

5.3) Bottom-up programming is the process of going from a lower level of organization, such as classes, to a higher level, such as program design.

5.4) OOP binds data and methods into a class definition and facilitates creating reusable code which can eventually save a lot of work.

5.5) C++'s OOP aspect has been inspired by a computer simulation language called Simula67.

5.6) The name C++ is derived from the C increment operator ++, that adds 1 to the value of a variable.

5.7) The OOP aspect of C++ gives the language the ability to relate to concepts involved in the problem while the C part of C++ gives the language the ability to get close to the hardware.

6)   Portability And Standards

6.1) C++ is portable in the sense that one can run it on a new platform without having to make any changes to the code one has written.

6.2) Hardware can be a problem to portability and a program which is hardware-specific is not likely to be portable.

6.3) Language divergence is also an obstacle to portability. (Hence, the importance of a published standard describing exactly how the language works - the ANSI standard for C++ has been developed.)

6.4) The ANSI C++ standard will draw upon the ANSI C standard as C++ is supposed to be, as far as possible, a superset of C.

7)   The Mechanics Of Creating A Program

7.1) After writing a C++ program, the following steps can be followed to get it running:-

a) Use some sort of text editor to write the program and save it in a file (which constitutes the source code).

b) Compile the source code (translate the source code to machine language - the file with the translated program is the object code).

c) Link the object code with additional code or code for other functions (to obtain a file containing the executable code which is the final product).

8) Creating The Source Code

8.1) Some C++ implementations, such as Turbo C++ and Zortech C++, provide an integrated environment which lets one manage all steps of program development, including editing, from one master screen.

8.2) In naming a source file, one needs to use the proper suffix to identify the file as being a C++ file. (This not only tells one the file is C++ source code, it tells the compiler that, too.)

8.3) The suffix is made up of a period followed by a character or group of characters called the extension.

8.4) The extension one uses depends on the C++ implementation, e.g., spiffy.c is a valid AT & T C++ source code file name.

9) Compilation And Linking

9.1) Initially, Stroustrup implemented C++ with a C++ - to C translator program, called cfront (for C front end), which translated C++ source code to C source code, that then could be compiled by a standard C compiler.

9.2) Often, the distinction between a translator and compiler is almost invisible to the user and the mechanics of compiling depend on the implementation.

10) Unix Compiling And Linking

10.1) A person on a UNIX system using the AT & T Version 2.0 C++ could compile his C++ source code file spiffy.c by typing the following command at the UNIX prompt:

```
cc spiffy.c
```

10.2) If the program has no errors, the compiler generates an object code file with an o extension.

10.3) To run the program, the name of the executable file, "a.out", should be typed.

10.4) If a new program is compiled, the new a.out executable file replaces the previous a.out.

10.5) As in C, in C++ one could spread a program over more than one file.

11) Turbo C++ and Borland C++ Compiling And Linking

11.1) In Turbo C++'s and Borland C++'s integrated environment, which has a built-in editor, a menu bar, accessible through a mouse or through Alt-key combinations, could be used to make one's desires known.

11.2) If one develops a program using more than one source code file, one could use the Project menu to open a new project.

11.3) Beside the manuals, both Borland C++ and Turbo C++ have a tutorial program which shows one the ins and outs of the Borland C++ or Turbo C++ environment.

12) Zortech C++ Compiling And Linking

12.1) Zortech C++, like Borland's Turbo C++, compiles one's C++ source program directly to object code rather than using an intermediate C source file.

12.2) For use with programs that require a large amount of memory, a command-line compiler is provided.

12.3) Zortech C++ includes a complete integrated environment, known as Zortech Work Bench (ZWB), which features menu bars, multiple movable and resizable windows, and full mouse support - the menus include File, Edit, Search, Window, Compile, Options, Browse and Help.

12.4) To develop a program in Zortech C++, one could use the Edit functions to create one's source file.

13) SETTING OUT TO C++

13.1) When one learns a computer language, one should start with learning the basic structure for a program; only then could one move on to the details such as objects and loops.

14) C++ Initiation

14.1) C++ is case-sensitive - in other words, it discriminates between uppercase characters and lowercase characters.

14.2) One has to be careful to use the same case as in the examples.

14.3) One constructs C++ programs from building blocks called functions.

14.4) A program is typically organized into major tasks and separate functions are then designed to handle those tasks.

15) The Main ( ) Function

15.1) The following lines state that there is a function called main ( ) and describe how the function behaves:-

```
int   main   (void)
{
     statements
     return 0;
}
```

15.2) They constitute a function definition.

15.3) A return statement, the final statement in main ( ), terminates the function.

16)   Statements And Semicolons

16.1) To understand one's source code, a compiler needs to know when one statement ends and another begins.

16.2) A statement is a complete instruction to a computer.

16.3) In Pascal, a semicolon is used to separate one statement from the next, though in certain cases it can be omitted, for example, just before an END, when one is not actually separating two statements.

16.4) But C++ (like C) uses a terminator rather than a separator (which is used by some languages to separate a statement).

16.5) Thus, the semicolon which marks the end of the statement is part of the statement rather than a marker between statements.

16.6) In C++, one could never omit the semicolon.

32 SIMPLE GUIDE TO SQL

1) Structured Query Language (SQL) has the ability to create and define relational database objects.
2) The "CREATE" statement brings out the Relational Database Management System (RDMS) objects, for example, Stogroup, Database, Tablespace, Table, Index, View, Synonym and Alias.
3) Four SQL data manipulation statements (DML) are available, namely, Insert, Select, Update and Delete.
4) These SQL statements perform RDMS operations which can affect only one row at a time, or, many or all of the rows in a table, as and when required.
5) Additional language requirements which provide the ability to process the table data while it is being retrieved and functions which modify the value of the data that is returned in a query are available.
6) SQL provides the ability to filter what data is retrieved in a select statement by including the WHERE clause, which specifies a variety of comparisons between two values.
7) SQL gives four arithmetic operations, namely, addition, subtraction, multiplication and division.
8) It provides the ability to summarize data as it is retrieved from a table via the GROUP BY clause.
9) It also provides the ability to sort the data retrieved from a table via the ORDER BY clause.
10) It also provides the ability to perform set manipulation operations, for example, one can SELECT the intersection of two or more sets of data by coding a JOIN.
11) Before one creates a table, one needs the following RDMS objects: an existing database and tablespace.
12) One has to first use the column name in the RDMS object, that is, the CREATE TABLE statement.
13) Columns can be added to a table after it has been defined by using the SQL ALTER TABLE statement.
14) New columns can be added to the end of the table.
15) Removing columns from an existing table involves a migration program which extracts only the desired columns of data, redefining the table without the unwanted columns, then populating the new table.
16) The SQL ALTER statement changes a table index, a table, a tablespace or a stogroup.
17) To create tables, one needs CREATETAB privileges.
18) To create tablespaces, one needs CREATETS privileges.
19) In order to enforce the uniqueness of the table's primary key a table index has to be created.
20) A synonym is an unqualified alternative name for a view or table.
21) A view is a virtual table which can represent all or part of one or more tables.
22) A view can present data which is the result of a JOIN or UNION of more than one table.
23) A foreign key is the key which is defined in one table to reference the primary key of a reference table.
24) Referential integrity refers to the automatic enforcement of referential constraints which exist between

a reference table and a referencing table.

25) The "cascade on delete" specification tells the RDMS to delete all dependent rows from the dependent table, at the same time honoring any delete rules which exist in those dependent tables, while the restrict rule tells the RDMS to fail the delete request if a dependent row exists.

26) The value of a primary key of a reference table can only be changed if it has no dependents.

27) The only local sub-system is the one where your application runs, while all other RDMS objects are remote.

28) On a remote sub-system, an SQL CREATE, ALTER or GRANT cannot be executed.

29) There are several things that one can do with an alias but cannot be carried out with a synonym, for example, one can retrieve data from remote tables and views while this cannot be achieved with synonym.

30) A SELECT statement is an SQL statement which retrieves data from a table or view.

31) Column-name qualifiers could be table names, view names, synonym names, alias names or correlation names.

32) A correlation name is a special type of column designator which connects specific columns in the various levels of a multilevel SQL query.

33) A correlation name can be defined in the first clause of an UPDATE or DELETE statement and in the FROM clause of a query.

34) A query which is written as part of another query's WHERE clause is known as a subquery.

35) A correlated subquery is one which has a correlation name as a table or view designator in the FROM clause of the outer query and the same correlation name as a qualifier of a search condition in the WHERE clause of the subquery.

36) The subquery in a correlated subquery is reevaluated for every row of the table or view named in the outer query; the subquery of a non-correlated subquery is evaluated only once.

37) A results table is the place that holds the results of a query.

38) To make a set of rows available to a program, a cursor, which is a named control structure, is used.

39) When the OPEN CURSOR statement is executed, the results table for a query specified in a DECLARE CURSOR statement of a cursor is created.

40) By issuing a FETCH cursor-name statement, data is brought into a program's working storage area from a results table.

41) A host variable is a variable which is referenced by embedded SQL.

42) An SQL error code of 100 appears when trying to fetch data from an empty or exhausted cursor.

43) The SQLA has to be specified in the Working Storage Section of a COBOL program by an SQL INCLUDE statement.

44) A positive value in the SQL CODE shows a successful execution, but with an exception condition.

45) A negative value shows that the SQL statement did not execute because of an error condition.

46) In dynamic SQL, the SQL statements can be changed, prepared and bounded by the program as it is

running; static SQL statements are prepared and bound before execution.

47) An SQ LDA cannot be specified in a COBOL program.

48) The SUBSTR function is a scalar function which breaks up a character or graphic string according to the substr argument list.

49) Column functions are features of SQL which allow one to calculate a single value derived from one or more values found in the specified column.

50) The only type of average one can calculate with the AVG column function is the mean average, which is calculated by dividing the sum of all values by the count of values.

51) The column function calculates its result based on the individual groups created by the GROUP BY specification.

52) A predicate is the SQL language element which specifies a condition about a value or set of values that may be true, false or unknown.

53) The different types of predicates are as follows:-

a) Basic predicates
b) Quantified predicate
c) Null predicate
d) In predicate
e) Between predicate
f) Like predicate
g) Exists predicate

54) The above-mentioned predicate types can be preceded by a "NOT" to reverse their meaning.

55) The table named in the From clause of a Delete statement containing a subquery is not the same table referenced by the subquery.

56) Addition and subtraction are the only arithmetic operations that can be applied to Date or Time data types.

57) The HAVING clause is coded after the GROUP BY clause in a query which is summarizing results by one or more grouping columns.

58) The output of a query can be sorted by using the ORDER BY clause and referring to the expression by its position in the list of columns named in the SELECT clause.

59) The results of two or more SELECT statements can be combined by using the UNION or UNION ALL keyword which causes the results of each SELECT statement to be combined as a single result table.

60) A Join depends on the common values of the columns named in a search condition; a UNION keyword would have to be used if there are no common values in two or more tables.

61) When merging the multiple result (one for each table in the union), the UNION keyword causes the

duplicate rows to be removed from the final result, while the UNION ALL keyword is processed to get all of the duplicate result rows to remain in the merged result table.

62) The ORDER BY clause is coded in the last SELECT statement in a union of multiple tables.

63) The results of any valid SELECT statement can be merged with the results of another.

64) A subquery cannot have a UNION keyword and only outer-level queries can use the UNION or UNION ALL keyword.

65) We can tell which table a result row came from by placing a different constant at the end of the column list in the SELECT clause of each SELECT statement in the union.

66) The WHERE CURRENT OF clause is used in cursor processing to execute an UPDATE or DELETE statement.

67) To use the UPDATE WHERE CURRENT OF clause during cursor processing, we have to code a FOR UPDATE OF clause listing all of the columns we wish to update.

68) We can make sure that the changes made to a table become permanent by issuing an SQL COMMIT statement.

69) A Unit of Recovery is the amount of processing in a program which is recoverable (can be undone).

70) When a ROLLBACK or COMMIT is issued, all of the open cursors are immediately closed and all cursor positions will be lost.

71) To continue processing the same cursor from the last position before the last COMMIT, we have to construct our cursor so that it uses the unique sort key (use the ORDER BY clause) of the result table being processed as a value compared by a "greater than" predicate.

72) The asterisk "*" allows us to select "ALL COLUMNS" from the table named in the SELECT statement without having to name the columns explicitly.

73) It is a bad practice to use the SELECT * in static SQL because if the table definitions for the tables used in a static program change (i.e., columns are added) all of the "INTO" clauses for those changed tables will become invalid, causing the programs to end abnormally.

74) A STOGROUP is a named object which lists the DASD volumes that are designated to store DB2 data.

75) The isolation level is specified in the BIND or REBIND statement.

76) With the isolation level set to Cursor Stability during read-only processing, a page lock is held all the while an applications cursor is positioned on it; with the isolation level set to Repeatable Read, all of the pages of data selected by the DECLARE CURSOR statement are locked until the next explicit or implicit commit.

77) A Smallint data type is 2 bytes long and has a range from -32768 to +32767, while an Integer is 4 bytes long and stores numeric data ranging from -2147483648 to +2147483647.

78) A solution table is another name for a results table and is a temporary table made to store the result of an SQL query.

79) The act of referring to a single table more than one time from a single SQL QUERY is known as a

recursive reference.

80) The UNION keyword cannot be placed within a SELECT statement; it is used to combine two or more SELECT statements.

81) The LIKE search condition allows one to select table rows based on a comparison of partial strings.

82) A "%" represents from 0 to many characters whereas a "_" represents a single unknown character.

83) We can select the lowest value from a numeric column by using the MIN column function.

84) We can get a string representation of a signed decimal number by using the DIGITS scalar function.

85) Authority is the privilege level required to access data from an RDMS and is needed all the time.

86) The CREATE TABLE statement or ALTER TABLE statement defines a foreign key.

87) A synonym is used for convenience – a synonym created for a table allows that table to be accessed by all privileged user IDs without having to use the Authorization ID as a qualifier.

88) To change the data in a table the application has to acquire an exclusive lock on the page on which the changing data resides.

89) The minimum lock level instituted for an enquiry is a SHARE lock.

90) The Delete command is restricted when the user does not have Delete privileges on the specified table or if there is a referential integrity constraint enforced for a dependent foreign key.

91) A searched update can update one or more rows of data depending on whether the rows satisfy the search conditions written into the update statement while a positioned update updates the single row pointed to by the cursor named in the update statement.

92) A positioned update cannot operate on a read-only cursor, e.g., one where the results table is made with a JOIN (results table made of data from more than one table).

93) We cannot work through a results table in reverse order, but we can specify the sort key as descending in the ORDER BY clause.

94) We cannot remove a column of data from a table without dropping the table, but, we can create a view and omit the column from the view.

95) When the volume is removed from a STOGROUP with the ALTER STOGROUP statement the table data on the volume remains available as before, but, the volume will not be used the next time DASD is allocated for that STOGROUP.

96) When the primary index of a table is dropped, nothing happens to the table data; only the index space is dropped.

97) The WHENEVER statement is used for specifying the HOST LANGUAGE statement to execute every time the specified exception condition exists.

98) The following exception conditions can be trapped by the WHENEVER statement:

NOT FOUND, SQLERROR,
and, SQLWARNING.

99) The scope of a WHENEVER statement is controlled by its placement in the listing and not by its execution order.

33 SIMPLE GUIDE TO DIGITAL COMMUNICATIONS

<u>Digital Communications System And Information Theory</u>
1) In his pioneering work, Claude Shannon formulated the basic problem of reliable transmission of information in statistical terms, using probabilistic models for information sources and communication channels, giving rise to a new field which is now called information theory.
2) Information theory is now widely applied in the design of digital communication systems.
3) The increase in the demand for data transmission, together with the development of more sophisticated integrated circuits, have now led to the development of very efficient and more reliable digital communications systems.
4) Shannon adopted a logarithmic measure in describing quantitatively the information content of a source.
5) <u>Model Of A Digital Communication System</u>
5.1) The information source generates messages that are to be transmitted to the receiver.
5.2) In a digital communications system, the messages which are produced by the source are usually converted into a sequence of binary digits (source encoding or data compression).
5.3) The sequence of binary digits from the source encoder is transmitted through a channel to the intended receiver.
5.4) Generally, no real channel is ideal, for the channels may have non-ideal frequency response characteristics and noise disturbances and other interference which corrupt the signal transmitted through the channel.
5.5) To overcome such noise and interference and, hence, to increase the reliability of the data transmitted through the channel, it is often necessary to introduce in a controlled manner some redundancy in the binary sequence from the source.
5.6) An option to providing added redundancy in the information sequence as a means of overcoming the channel disturbances is to increase the power in the transmitted signal (important to design power efficient communications systems as transmitter power is often very expensive).
5.7) Each bit from the channel encoder is transmitted separately (binary modulation).
5.8) At the communications system's receiving end, the digital demodulator processes the channel-corrupted transmitted waveform and reduces each wave-form to a single number which represents an estimate of the transmitted data symbol (binary or M-ary).
5.9) The demodulator must decide which of the M wave-forms was transmitted in any given time interval when there is no redundancy in the transmitted information.
5.10) The frequency with which errors occur in the decoded sequence is a measure of how well the demodulator and the encoder perform.
5.11) When an analog output is needed, the source decoder accepts the output sequence from the channel decoder, and, from knowledge of the source encoding method used, attempts to reconstruct the

original signal from the source.

6)   Logarithmic Measure For Information

6.1) We have to select an appropriate measure for information.

6.2) We have to consider whether two discrete random variables (X and Y) are statistically independent or dependent.

6.3) A high-probability event conveys less information than a low-probability event.

6.4) The logarithm of the ratio of the conditional probability is:

$$P(X = x_i \mid Y = y_j) \equiv P(x_i \mid y_j)$$

6.5) The logarithmic measure for information content is the appropriate one for digital communications.

7)   Average Mutual Information And Entropy

7.1) The average value of the mutual information can be obtained by simply weighing $I(x_i; y_j)$ by the probability of occurrence of the joint event and summing over all possible joint events.

7.2) Therefore, we obtain:

$$I(X;Y) = \sum_{i=1}^{n} \sum_{j=1}^{m} P(x_i, y_j) I(x_i, y_j)$$

$$= \sum_{i=1}^{n} \sum_{j=1}^{m} P(x_i, y_j) \log [P(x_i, y_j)/ P(x_i) P(y_j)]$$

as the average mutual information between X and Y.

8)   Information Measures For Continuous Random Variables

8.1) The definition of mutual information given for discrete random variables may be extended in a straightforward manner to continuous random variables.

8.2) However, a continuous random variable requires an infinite number of binary digits to represent it exactly. (Therefore, its self-information is infinite and, hence, its entropy is also infinite.)

9)   Sources, Source Models And Source Coding

9.1) By comparing the average number of binary digits per output letter from the source to the entropy H (X), a measure of how efficient a source encoding method is can be obtained.

9.2) The encoding of a discrete source having a finite alphabet size may seem a relatively simple problem, but this is true only when the source is memoryless, i.e., when successive symbols from the source are statistically independent and each symbol is encoded separately.

9.3) On the contrary, a discrete-time source whose output is characterized by discrete random variable X, a continuous-time source, that is also called an analog source or a wave-form source, has an output X (t) which is characterized as a stochastic process. (The first step in the encoding of the output of such a source usually involves sampling the source output periodically.)

9.4) Quantizing each sample separately is known as scalar quantization.

9.5) Quantizing a block of samples jointly as a single vector is called vector quantization or block quantization.

34 SUCCESSFUL AUTOMATION

The advantages of automation are obvious but not every company should automate. Automation involves finance and the financial aspects should be carefully considered first before embarking on it. It would be foolhardy to "follow the Joneses" when it comes to automation. There is presently a big hoo-hah regarding automation. But it is always wise to bear in mind that even low cost automation may cost thousands of dollars, which even if a small percentage of which were offered as productivity incentive to workers, may help improved productivity significantly.

What does automating the assembly line or the work-floor involve? Automation is the process of using electro-mechanical devices, in replacement of human labor, to perform jobs in the factory, especially work which is dirty, monotonous, dangerous and hazardous to the health of workers. It involves the use of robots, computers, "clever" mechanical devices, or other components/parts such as proximity switches, sensors, cams and cam followers, linear motion bearings, inner races, programmable logic controllers (PLCs) and the like. Some of the brand names in the market for components/parts for automation are Omron, Sun-X, Keyence, NTN, Koyo, Pepper + Fuchs and Hitachi. Some of the the makes of systems, or robots, for automation are Robert Bosch, Asea, Fanuc, CRS, GMF, Unimate and Odex.

In investing in automation, the following financial aspects should be considered with great care first:-

(1) Pay-back. This is the most popular method of financial analysis as it is easy to understand and apply. Here, we have to estimate the time taken for all the various incomes resulting from a project to equal (pay-back) the original expenditure. That is, the cash break-even point has to be measured. The investment has to be recovered before the equipment becomes technically obsolete or loses its working life. As a rule, the pay-back should not exceed three years. The higher the risk the shorter the pay-back should be.

(2) Return On Investment (ROI). This is normally considered to be the percentage/ratio of annual profit from a project divided by capital cost. Different organizations use various combinations of tax, interest charges and depreciation to come to a figure for the profit. There are no hard and fast rules. The disadvantage of this method of looking at an automation project is that it does not reflect the useful life of a project or consider the timing of cash flows, and the "profit figure" can be manipulated by simply changing the depreciation policy which is generally arbitrary.

(3) Net Present Value (NPV). Unlike the earlier two methods, here, the "time value" of money is taken into account. A given sum of money is worth more to a company now than it is going to be in the future, because in the interim period the money can be put to profitable use by earning interest in the bank as fixed deposit. The present value, P, of a sum of money, S, arising in n years is thus obtained

by the following formula: $P = S \times 1/(1 + r)n$, where r is the rate of bank interest per annum S can earn. The value of any cash flows that will arise in the future can be discounted by using the appropriate discounting factor $1/(1 + r)n$ for the year in question to obtain their present equivalent values. The automation project is thus viewed in terms of a series of equivalent present values. It is therefore possible to compare legitimately cash flows which will occur in five years with cash flows that occur now, using such discounted cash flows (DCF) for all of the cash flows in the project as they have been translated into their equivalent "present-day values". The net present value (NPV) for the whole project is obtained by adding together all the various discounted cash flows (with negative values corresponding to outgoings and positive to earnings). The NPV will indicate whether the project is feasible or not. If the NPV is zero the project just breaks even in "present value" terms, which means that the company neither loses nor gains from investing in the project, and it will be better to keep the money in the bank to earn interest. If the NPV is greater than zero, the organization will enjoy a net gain if it undertakes the project, whereas if it is negative, the firm will be better off leaving the money in the bank (where it will earn interest). However, this method of financial evaluation may not be suitable as it seeks to look at things in the "short term", whereas automation projects tend to take a relatively long time to bear fruits. Yet it can be considered ideal when a comprehensive approach is used by including both direct and indirect factors over the long run.

(4) Internal Rate Of Return (IRR). Discount rates which are successively higher are applied to a project (that is, assume bank rates to be progressively higher, e.g., 10 %, 15 %, 20 % and so on) until a rate is found at which the project appears to break even, with a NPV of zero. The final value obtained for the interest rate (say 30 %) is considered the internal rate of return (IRR) for the sum invested in the project. That is, investing in the project is equivalent to investing the money in a bank which is offering an interest rate of (30 %). The IRR calculations concentrate on analysing the profitability of predicted incremental cash flows. For a project to be feasible, its IRR has to be higher than the company's cost of capital. However, interest rates fluctuate from time to time. To overcome this set-back, the technique involves varying to reasonable degrees different parameters, one at a time, to find out the sensitivity of the project's IRR (and NPV) to the variations. This approach may be cumbersome but it is worthwhile. The IRR is conceptually easily understood. Its computation is therefore desirable.

(5) Leasing. This is the alternative to purchasing automatic equipment and devices. Some automatic equipment makers and robot manufacturers do lease out their products. Most banks and leading financial institutions provide leasing facilities. The lessee remits a regular sum which is adjusted to take account of the capital allowances which the lessor is able to claim on the equipment leased. Normally no deposit or down-payment is needed. Often, payments can be varied to suit a business

cash-flow position. The lessee is normally entitled to lease the equipment at nominal rent only, at the end of the initial leasing period. In leasing, the lessor actually deals with depreciation, cost of capital, and other related matters; the lessee is spared such "headaches". Normally, the lease lasts for five to seven years.

(6) Other costs have also to be considered such as cost of special tooling, installation cost, maintenance costs, operating costs, programming costs, cost of capital, and depreciation. After all, it may not be feasible to automate. Perhaps, replacing existing tools and equipment with more up-to-date, efficient ones, implementing attractive productivity incentive schemes, training and educating the work-force, introducing special jigs and fixtures, improving work methods or the simple expedient act of supplanting idle, inefficient workers with good ones may perform the "trick". Selecting the technology for automation is another area to look into. Sometimes, it is not feasible to automate some aspects of the factory. Should robots be used? Should it be low cost automation? Should any mechanical devices which are automatic be utilized at all? How about installing flexible manufacturing cells or computer integrated manufacturing systems? How about auto-insertion equipment (for electronics assembly)? And so forth. There are many systems designers, and, the value of the IRR (sometimes known as yield or discounted cash flow rate of return on investment) enables projects to be compared directly against the interest rates which are obtainable from the banks. The following are some of the systems designers:-

(1) Unimation, Inc.
(2) Cincinnati Milacron
(3) Prab Robots
(4) Machine Intelligence Corporation
(5) Androbot, Inc.
(6) Heath Electronics
(7) Rhino Robots, Inc.
(8) Microbot, Inc.
(9) Air Technical Industries
(10) IBM Corporation
(11) Odetics Inc.
(12) Remote Technology Corp
(13) Robot Systems International
(14) Cybermation
(15) Hitach Ltd.
(16) Inspectronics
(17) Rockwell International
(18) Pentek Inc.

(19) Sumitomo
(20) Westinghouse Hanford
(21) Robert Bosch GmbH
(22) G.E.C. Australia Ltd.
(23) Euram Packaging Systems Pte. Ltd.
(24) The Commonwealth Industrial Gases Ltd.
(25) The Lincoln Electric Company (Australia) Pte. Ltd.
(26) Liebherr Verzahntechnik GmbH
(27) NBS

Technical information can normally be obtained from these companies, free of charge. Engineers from these systems suppliers are only too willing to study your factory organization and offer systems proposals. Your factory engineers should be able to have free discussions with the suppliers. A suitable system could be selected and installed in a matter of time.

Sometimes, a budget approval has to be obtained from the board of directors. A system has to be within the approved budget. Moreover, the staff have to feel comfortable with the technology involved and must be prepared to master it. Advanced systems like the Pana Robot of The Commonwealth Industrial Gases Ltd. is flexible, easily programmed and suitable for diverse welding tasks, the Euram Robot of Euram Packaging Systems Pte. Ltd. is utilized for palletising and packaging purposes, and the Robert Bosch flexible assembly system, which is a modular system designed to master future manufacturing requirements, which utilizes state-of-the-art techniques for such sectors as feed technology and handling technology, and which is capable of handling a wide variety of work-pieces, are now available in the market. For those hooked on automation, proximity switches, sensors, pressure switches, stepper motors, induction motors, belt drives, pulleys, relays, timers, limit switches, pneumatic devices, hydraulic systems and sequence controllers could be used. For instance, servo motor driver linear actuators could be used to handle parts and components, gravity feeds could be used to sort components, an inclined vibrating shooter could be used to transfer semi assembled parts from one point to another, fibre optic sensors could be used to "cut off" the motor when the raw material runs out, limit switches could be used whereby if, e.g., the flywheel guard were not lowered (for the sake of the worker's safety), the limit switch would not be "on" and the machine would not run, and so on, depending on the ingenuity of the engineer or system designer. Sequencers or programmable logic controllers (PLCs) using adder logic are normally utilized to time and control the sequence of work flow, independently of the worker who just keeps an eye on things and resets the equipment should there be stoppages or work-pieces jams; the PLCs "instruct" relays (switches) to go "on" or "of" in sequence and at regular intervals, enabling various sections of the machine to perform certain actions or halt at regular intervals.

In the more advanced systems, such as the computer integrated manufacturing system (CIM), all the various departments in the organization are linked by a computer network, and even inter-company links are possible; in the case of a flexible manufacturing system (FMS), a distinct advantage is that there is product flexibility, e.g., shoes may be produced when the demand is high and if it is no more profitable to make shoes, it is relatively easy to switch over to another product line, say tooth-brushes, pens, or folders.

There has been some debate on what is considered automation. The S'pore Robotics And Automation Association (SRAA) Project Team in a survey some time ago considered automation to mean equipment falling under the following categories:-

(1) Special purpose automatic machines
(2) Standard production machines
(3) Materials handling and parts feeding equipment
(4) Programmable process control equipment
(5) Automatic inspection and measuring equipment
(6) CNC machine tools
(7) Sequential, pick and place industrial robots
(8) Computer-aided design and manufacturing systems
(9) Re-programmable industrial robots
(10) Automatic-warehouses, computerised storage and retrieval systems
(11) Automatic guided vehicles

Japanese companies seem particularly strong on automation. For example, a local subsidiary of a well-known Japanese multinational corporation which manufactures component parts for the compact disc player utilizes advanced automated assembly systems and ninety per cent of its manufacturing process is automated. A newly set up Japanese factory here operates completely unmanned at night after everyone in the factory has retired for the day. What a saving in labor costs! Perhaps, the reason why the Japanese are so amenable to robots and automated systems is because they do not feel threatened by such systems as they have been "promised" life-long employment. This is not really so with the others. Automation always, or almost always, brings about the fear of loss of jobs or retrenchment. Of course, this need not be the case, as existing staff can be retrained and be re-deployed to more challenging, interesting work.

Yet some still prefer to stick to their old work habits and abhor any changes, even changes for the better. So, implementing an automated system will not be entirely smooth sailing for some.

At the moment, automated systems are already in use in the following industrial sectors here:-

(1) Electrical and electronics
(2) Precision engineering and metal working
(3) Shipbuilding and ship-repair
(4) Tool, die and mould making
(5) Food and beverage
(6) Chemicals and chemical products
(7) Wood and wood products, including furniture
(8) Textile, apparel and leather products
(9) Paper and paper products
(10) Printing
(11) Plastics
(12) Warehousing and materials handling
(13) Others

Companies here are encouraged to automate by the government as there is an acute labor shortage problem. They appear much too dependant on foreign workers, especially Malaysians. Despite the hefty levies for employing foreign workers which now stand, e.g., at $300 per head per month for Malaysian factory and $350 per head per month for Malaysian construction and construction related workers, the engagement of foreign workers has not abated to any great extent. The plausible solution to this problem appears to be more automation. The Economic Development Board here has the following incentive schemes for automation:-

(1) Automation Feasibility Study Scheme (AFS)
(2) Training Grants For Automation Leasing Scheme
(3) Investment Allowance Scheme
(4) Design For Automation Scheme

The Board provides training for automation manpower at its ASEA-EDB Robotic Training Unit (AERTU). It provides computer-based courses at its Micro-cadd Laboratory (JSTI) and Mentor Graphics-EDB CAE (IC Design) Training Unit (MECTU). It provides consultancy and feasibility studies whereby its consultants can be attached for two weeks at a company to carry out the study. Another set-up, SAL Industrial Leasing Pte. Ltd., provides incentive for automation in the form of the low-interest loan and attractive terms.

In carrying out the automation project, some of the following problems are likely to be encountered:-

(1) Heavy investment and financial burden

(2) Risks and uncertainties
(3) Initial high operating cost
(4) Inadequate in-house knowledge, experience and skills
(5) Lack of confidence in automation equipment suppliers
(6) Inadequate service/expertise rendered by vendor or supplier
(7) Lack of training opportunities and qualified trainers
(8) Lack of financial backing
(9) Problem of integrating new and old system
(10) Problem of displaced workers and antipathy
(11) Lack of technical expertise within the company to select or specify the right automation system
(12) Lack of motivational techniques for the new technology
(13) Others such as lack of software services and maintenance capability

On the other hand, improvements in the following areas are likely to result from a successful automation exercise:-

(1) Reduction in labor costs, possibly resulting in unmanned operations even which will greatly help relieve any serious labor shortage problem
(2) Improvement in product quality
(3) Quicker response to changes in demand
(4) Reduction of rejects/wastage
(5) Reduction of delivery lead-time
(6) Increased manufacturing capacity
(7) Higher profits
(8) Better inventory control
(9) Reduction in machine down-time
(10) Better design capability
(11) Better safety
(12) Higher operating ratio and efficiency improvement
(13) Flexibility - options of add-on ancillary parts and inter-faces with other equipment are possible
(14) Capability of handling a wide variety of work-pieces

It is generally believed that the advantages of automation far outweigh the disadvantages. Nevertheless, it is never wise to rush into it without thorough planning. The following are proposed procedures on going about implementing the automated system after the "green light" to go ahead has been given:-

(1) Initiate

(2) Get everyone into initial discussions, including labor force representatives such as shop stewards and union leaders
(3) Determine the medium and long-term objectives
(4) Get agreement on priorities
(5) Carry out detailed planning of approach for evaluation if possible
(6) Establish the targets
(7) Set the schedule
(8) Consolidate expertise on automation and robotics
(9) Look into existing solutions
(10) Look into suitable applications in detail
(11) Consider alternatives to automation
(12) See whether reorganization is necessary
(13) Look into product design alterations
(14) Determine automated system requirements
(15) Determine ancillary equipment requirements
(16) Determine manpower requirements
(17) Consider the likely effects of automation
(18) Determine the appropriate approach to purchase
(19) Carry out a detailed cost analysis
(20) Finalise the detailed specification of the automated system
(21) Arrange for the comprehensive training of staff
(22) Look into a maintenance program for the automated system
(23) Look into the safety aspects of the system
(24) Implement the system
(25) Constant monitoring after implementation

When the system is implemented, it is quite normal for the system supplier, at least initially, to provide some kind of back-up support or after-sales service for the equipment as there are bound to be teething problems. But, the organization cannot expect this to last forever and must ensure that its own staff are eventually able to run the system independently. Staff training is therefore of great importance; maintenance personnel should be given sufficient training so that they are capable of carrying out preventive and break-down maintenance work.

Finally, a few important words of advice for the organization, especially those who feel that they are going to lose their jobs as a result of automation. It is of vital importance to educate all within the company to see the fact that automation benefits them in many ways, such as getting rid of dirty, monotonous, even dangerous, jobs, more challenging, higher level work (which may mean pay increase or promotion) and higher profits for the company which mean higher year-end bonuses for the employees. Employees must never be made to feel that automation is being implemented to cut cost at their expense. It is important that there is no antipathy from any of the staff. As such, employees, especially those in important positions, e.g., the front-line management staff such as supervisors and foremen, have a very important role to play. They have to be convinced first of the benefits and advantages of automation and will play a significant part in educating the workers who report directly to them. It is well and good to automate and be at the cutting edge of technology but do not let yourself be cut and left bleeding.

35 PRODUCTION PLANNING AND CONTROL

In production planning and control, two important objectives have to be borne in mind and attained, namely, producing the required quantities of a given product and producing these quantities at appropriate times. However, the system of planning and control adopted and the procedures in involved will vary from organization to organization, e.g., that in a firm with data-processing equipment will differ somewhat from that in a firm which does not have such equipment. It should be noted that regardless of how elaborately a system is developed, things will not work out perfectly at all times. Operation schedules developed by the most scientific methods will be disrupted because an employee decides to take the day off or because a supplier addresses a shipment incorrectly or because someone inadvertently jams his tool into the working parts of a machine. Furthermore, planning and control activities involve the use of judgement or are based on predictions of the future course of events, and errors in judgment and in forecasting are commonplace.

Moreover, in some companies, production planning and control are undertaken by the production control department, whereas in others, the function is part of the job of the manufacturing personnel. But, whether it is necessary to have a production control department or not does not matter, and there are firms which produce goods efficiently despite the absence of a production control department.

However, I would like to caution practitioners against excessive formality and centralisation in production control.

THE PRODUCTION PLANNER AND HIS PROBLEMS

Introduction
As mentioned earlier on, the production control department is the nerve centre of a factory. The production planner keeps in close touch with the staff of the production floor, monitors the progress of production and re-schedules the production as and when necessary. His job is not an easy one. He has to keep track of many things all at once. If he forgets or misses a few important things, chaos can result. For example, if he forgets to instruct the die-maker to prepare dies for a new model scheduled for production soon, the new model will not be able to commence production on time. The production planner has to be methodical in his work; he has to have a head for details if he were to perform well and to be able remain calm under pressure. Pressure there is bound to be. What with last minute orders from customers and delays in production work due to machine break-down or power failure or even labor shortage, shortage of raw materials, and other unforeseen disruptions.

Functions
The production planner normally performs several important functions. One of these is inventory control.

He estimates the material requirements at any one time and ensures that there is sufficient stock of them and no overstocking which will result in higher storage and material handling costs. Knowledge of inventory control methods such as JIT (Just-In-Time) may be useful. If he were involved in MRP I or MRP II, he has to have some computer knowledge. He also has to liaise with the purchasers regarding what, when and how much materials to buy.

Secondly, the planner monitors the progress of production, e.g., the status of work in process, finished goods, and so on. He has to re-schedule production work when production targets are not met.

Thirdly, there comes the actual work of production scheduling itself. He has to have knowledge of the production flow (probably, if he is new to the job, he needs to consult a production flow chart or even a quality control flow chart). He has to have product/technical knowledge. He may already have acquired some technical knowledge from college or university. But in most cases he will still require on-the-job training, especially on the technical aspects. He may be expected to understand blue-prints or schematics. He has to have an aptitude for figures as his work involves mainly figures.

Fourthly, though this is basically the task of the sales department, he may have to prepare sales forecasts. He may even have to prepare the master schedule for the whole factory though this is usually carried out by a more senior person, such as the production control manager. Of course, his basic job is to schedule production for each machine or section.

Fifthly, he schedules deliveries to customers ensuring that right quantities and the right items are delivered according to schedule. He may even have to liaise with customers regarding such matters. To perform the task of scheduling he has to have data on time standards. In other words, he has to know the time it takes to perform the various sub-operations, and the time, of course, it takes to perform a complete operation. Information of time taken to perform various jobs is normally gathered by work study officers or industrial engineers who carry out what we call "work measurement" and give advice on methods improvement. The production planner relies on his judgement and experience when scheduling production. At times, he has to consult the line supervisors directly responsible for production to find out whether schedules are realistic, achievable or not. Scheduling is very much a "numbers game". If the planner detests figures, production planning is not the work for him. Depending on how complicated the product-to-be-produced is, the task of scheduling requires patience, an eye for details, and thoroughness. As production normally has to proceed without delay or stoppage, the production planner has to work fast, to re-schedule production which has fallen behind schedules, or to advise on "change of items" as when there is a sudden cancellation or delay of orders or delivery. Technical knowledge and knowledge of production processes are essential.

Production Scheduling

The purpose of a production plan is to balance requirements (orders in hand/sales forecasts) against resources (people, materials, machines) so that the company operates at maximum efficiency and profitability.

The first step required when formulating a production schedule is to sort the orders or forecasts received into a sequence, according to the dates by which they are required. The orders have then to be broken down to establish what components need to be bought or made to complete the order.

From this, the production planner can determine what materials, machines and labor will be needed and the earliest possible date for final production. This is a crucial process, as although there may be standard lead times, circumstances may change from day to day, even hour to hour, e.g., being short of just one component can delay the whole order, leaving the factory with excessive levels of unfinished work in progress.

It is essential to consider the total factory loading, especially if the company makes a wide range of products. Unless each line is loaded evenly, one part of the factory may be working flat out, paying premium rates for overtime, while in other areas people and machines may be standing idle. In most manufacturing processes, there is a period of non-productive time when orders are changed over. Requirements for the same or similar parts should be batched together to minimise down time in order to achieve greater output. This balancing process should result in a schedule which can produce goods at a profit and satisfy the customer by the efficient use of available resources. There will be times when orders either cannot be produced on time, or only at an unacceptable cost. In such cases the customers should be informed. It is better to lose a single order but keep a good customer than to gain a reputation for unreliable delivery promises and thus lose customers in the long run. Achieving the best balance of resources while at the same time satisfying customer demands is a far from easy task.

In any production process, it is usually impossible to state precisely what can be produced with the given equipment, manpower and materials. Only the industrial giants can afford computer power to formulate the best production plan, according to a wide range of possible allocations against a sequenced order plan. Most companies have to rely on simpler systems, coupled with the flair and experience of their production planners. As computerisation becomes cheaper and more powerful, companies should regularly consider whether they could benefit from the mini- and micro-based systems appearing in the market.

Production scheduling is simpler if machinery is highly specialised rather than multi-purpose. Within the priorities of the order plan, production planners should ensure that all machines are loaded to the maximum possible and that down-time (due to, e.g., order change over or gaps in the production sequence) is kept to the minimum. When planning material availability, the production planner should ask the following questions:

(1) What material is required and when?
(2) Is the material available in stock at the moment?
(3) Can the material be made available in time for planned start of production at an acceptable cost?

Making use of available labor to the maximum is a complex factor which the production planner has to take into consideration. Particular equipment may require specialist skills which only a few workers prossess. There may be a long holiday period. It may be a time of the year when absence through sickness is traditionally high. The work-force may also resist changes in shift patterns or levels of overtime working. There may even be serious labor shortage. Industrial health may be poor resulting in "go-slows" or strikes.

To overcome such problems, it may be necessary to sub-contract work out. Other companies operating in the same industry may have spare machines and manpower capacity, or may have the right materials available. Reversing the process, i.e., taking on sub-contracted work, can also be a useful way of levelling out workload troughs. Overtime work may be another solution.

Balancing all the variations in resources against demand should result in a good production plan which would see to it that customer orders are fulfilled on time. It has to be borne in mind here that production scheduling is an individualistic task which depends on the planner's own experience, judgement, and creativity. Different production planners schedule a similar task differently, i.e., no two production plans are alike.

Production Control
A production plan is rarely, if ever, carried out in all its details. Machines break down, suppliers do not deliver on time, and workers take holidays, leave the company or go on strike. It is the task of the production controller to ensure that production is maintained in line with the production plan wherever possible. The production controller has to respond to the things which do go wrong and revise the plan in order to get back on schedule.

The production controller monitors the supply and production process to ensure that there is all the time accurate up-to-date information on what performance has been achieved and what actions are required to maintain performance. Where deviations from the production plan are spotted, corrective action should be taken at an early stage to overcome the shortfalls. The production controller should make regular checks on material availability and not just on the day it is due. This is because a supplier's promise given some time ago may not be kept. Such routine follow-ups should be backed by regular stock checks.

The production controller should be constantly aware of what labor hours are available, how the operation is performing in terms of quality and productivity, and even what social events are taking place, e.g., planned overtime work may be suddenly cancelled when a big sporting event is unexpectedly rescheduled.

If machines break down, outside contractors can be brought in to repair them quickly. On the other hand, machines can be hired or the job may be subcontracted for completion. Materials not available from one source can be bought from other sources. Alternatively, other materials can be substituted. Defects can sometimes be rectified. Design engineers may agree to the use of off-standard components if they do not materially alter the finished product. Manpower problems may be resolved by resorting to extra over-time work, additional shifts and engaging temporary labor. Customers can often accept a later delivery date, or accept part shipment of an order, if treated correctly. The production controller should always make customers aware of the status of their orders.

The function of production control should be the continuous process of checking, reworking and rescheduling. The production controller should see to it that the original production plan is adhered to as closely as possible though this is vastly difficult - changes should only be made if they were really unavoidable.

TIPS ON PRODUCTION PLANNING AND CONTROL

Important Elements Of An Order Plan
(1) Sequencing orders according to dates desired by customers
(2) Breaking down orders into details
(3) Maintaining balance between product lines
(4) Grouping orders together so that there are economies of scale
(5) Maintaining a balance between "high profit" and "low profit" orders
(6) Informing the customer of delays

Salient Points Of Production Scheduling
(1) Planning of materials according to the priority of the order
(2) Minimising of machine break-downs
(3) Accurate recording of stock levels
(4) Maximum usage of available manpower
(5) Subcontracting out for the purpose of smoothing work load peaks and troughs

Pointers On How To Balance Resources And Customer Needs In A Production Plan
(1) Look at orders from customers
(2) Look at sales forecast
(3) Look at stock level or subassemblies
(4) Consider the details
(5) Plan according to dates required
(6) Have customer satisfaction in mind
(7) Identify special requirements
(8) Aim for even distribution of workload
(9) Plan for batches of similar groups
(10) Consider the machinery required
(11) Consider whether manpower is sufficient and whether the required skills are available
(12) Consider subcontracting of work where necessary
(13) Consider the materials required, when they are required, whether the stock level is sufficient, and whether they can be procured in time
(14) Aim for a balance of "high profit" and "low profit" orders
(15) Consider effect of premium payments on profitability
(16) Look into downtime costs from excessive charges
(17) Look into the cost of subcontracting work

Problems That Can Affect Production
(1) Production quality that is substandard
(2) Low productivity
(3) Machine breakdowns
(4) Unforseen raw material shortages
(5) Inferior products which are rejected by customers
(6) Strikes at supplier's plants which affect delivery of materials
(7) Breakdowns in transport
(8) Bans on over-time work
(9) High absentee rate
(10) Substandard raw materials

POSSIBLE SOLUTIONS TO PRODUCTION PROBLEMS
(1) Rework substandard products. Effect changes through engineering design.
(2) Implement productivity bonus schemes or incentive schemes.
(3) Enter into machine servicing contracts. Plan and implement preventive maintenance. Hire machines.
(4) Use substitute raw materials. Order materials from another supplier who can supply them. Modify or

rework some of the parts.

(5) Modify or rework product. Change its design. Give customers improved products.
(6) Keep buffer stocks. Use other suppliers.
(7) Use alternative modes of transport, e.g., airing instead of shipping.
(8) Recruit more workers. Use temporary labor. Improve labor relations.
(9) Get workers to work over-time. Introduce additional shifts. Improve industrial relations.
(10) Complain to suppliers about substandard quality and ask for improvement and/or replacement. Alternatively, change suppliers.

CASE STUDIES
Case No. 1

In early 1982, JCB, the UK earth-moving equipment manufacturer, was faced with a total shut-down of their highest volume production line as a result of a protracted strike at the factory of their only engine supplier. If you had been the manager of JCB, what would you have done to overcome the problem of lack of material? (see solution below.)

Case No. 2

Ford Motor Company in Dagenham, England, which was at the time producing 1,000 vehicles a day, was threatened with a complete stoppage of work at their Cortina assembly line, due to a shortage of steering wheels which prevented vehicles being driven off the line. As manager of Ford Motor Company what line of action would you have taken? (See solution below.)

SOLUTIONS
Solution To Case No. 1

JCB resorted to the buffer stock they had held for cases of emergency. They had also put in a crash engineering development plan to redesign the unit to use an engine from an alternative supplier. The quick redesign combined with an extensive material procurement exercise enabled JCB to fit the new engine and thus maintain the production of their equipment.

Solution To Case No. 2

A squad of mechanics was used to drive the cars into the park and remove the steering wheels. The steering wheels were then used as "slaves" to drive off more cars. In this way, production was able to go on without stopping. Fitting new steering wheels when they became available was a relatively simple exercise.

PRODUCTION SCHEDULES (AN EXAMPLE)
PRODUCTION RATE PER WORKER (A CONSERVATIVE ESTIMATE)

1) 4-Key Pad:	5 Pieces Per Day
2) 8-Key Pad:	5 Pieces Per Day
3) Master Unit:	3 Pieces Per Day
4) Relay:	5 Pieces Per Day
5) Dimmer:	5 Pieces Per Day

WEEKLY PRODUCTION SCHEDULE FOR TWO WORKERS

MONDAY:	10 Pieces of 4-Key Pad
TUESDAY:	10 Pieces of 4-Key Pad
WEDNESDAY:	6 Pieces of Master Unit
THURSDAY:	10 Pieces of Relay
FRIDAY:	4 Pieces of Dimmer
SATURDAY (HALF DAY):	QC Test for Monday to Friday's Output

NOTE:- TO PRODUCE THE FOLLOWING ITEMS WITHIN THREE MONTHS (13 WEEKS), WE NEED MORE THAN TWO WORKERS:

1) 4-Key Pad:	350 Pieces
2) 8-Key Pad:	150 Pieces
3) Master Unit:	100 Pieces
4) Relay:	50 Pieces
5) Dimmer:	100 Pieces

We have thus to produce at least the following quantities in a week:

MONDAY :	27 pieces of 4-Key Pad
TUESDAY:	12 pieces of 8-Key Pad
WEDNESDAY:	8 Pieces of Master Unit
THURSDAY:	4 Pieces of Relay
FRIDAY:	8 Pieces of Dimmer
SATURDAY (HALF DAY):	QC Test

36 MANUFACTURING RESOURCE PLANNING: MRP II

INTRODUCTION

MRP has now expanded to mean more than material requirements planning. We now have "MRP II", which represents manufacturing resource planning, a system for planning and controlling the operational, engineering, and financial resources of a manufacturing firm.

FOUR CLASSES OF MRP USERS

In his book on manufacturing resource planning, Oliver Wright defines four classes of MRP users. The characteristics of the four classes are listed in the table below. The lowest level is Class D, in which the potential utility of material requirements planning is hardly realised at all. MRP is used basically as a data-processing system with many of the traditional production control procedures (e.g., shortage lists, and expediting) still being used. At the top of the list is the Class A MRP user. This is a company that uses material requirements planning together with capacity planning, shop floor control, and other components of a computer-integrated production management system (CIPMS), which as the name implies involves the use of the computer, a powerful tool, to help accomplish the vast data processing and routine decision-making chores in production planning that had previously been done by human beings. The next step beyond the Class A user is MRP II. To describe MRP II, we shall first examine the progressive evolution of material requirements planning into manufacturing resource planning.

Class Of User	Characteristics
Class A	Uses closed-loop MRP.
	Integrated system has MRP, capacity planning, shop floor control, vendor scheduling, etc.
	MRP system used to help plan sales, engineering, production, purchasing, etc.
	No shortage lists to over-ride the production schedules.
Class B	System has MRP, capacity planning, shop floor control, but no vendor scheduling.
	Not used much to help manage the business; used as a production control system.
	Needs help from shortage list.
	Inventory is higher than need be.
Class C	System used for inventory ordering rather than scheduling.
	Scheduling by shortage list.
	Master schedule is overloaded.
Class D	MRP working in the data processing department only.
	Inventory records are poor.
	Master schedule, if it exists at all, is overstated and mismanaged.

Relies on shortage list and expediting rather than MRP.

THE FOUR STEPS OF MRP

Material requirements planning has changed significantly over the years. The following four steps can be identified in the evolution of MRP:

(1) An improved ordering method
(2) Priority planning
(3) Closed-loop MRP
(4) MRP II

The first step was implemented when initial use of the computer was made to perform the requirements planning calculations. Before the computer, this task was performed manually and consumed tremendous amounts of time and manpower to accomplish. Computer-based MRP systems represented a tremendous improvement in the ordering of raw materials and components because of the speed and accuracy with which the requirements planning task can be performed.

The need for the second step in MRP evolution grew out of the attempts to implement step one MRP in conjunction with an unrealistic master schedule. This was a master schedule that ignored the limitation imposed by plant capacity and other constraints. It caused the MRP processor to generate schedules and requirements that could not be accomplished by the factory. As a result, the use of shortage lists continued. To overcome these problems, the MRP systems material requirements can be phased into time periods (weeks, or even days).

Priority planning not only provides a means for dealing with rush jobs by increasing their priorities, it also helps to unexpedite jobs whose priorities have been reduced.

Step three MRP represents the level of achievement of the Class A MRP user. Closed-loop MRP is an improvement over step two MRP because it not only plans the priorities but also provides feedback information relating to executing the priority plan. Closed-loop MRP means that the various functions in production planning and control (capacity planning, inventory management, shop floor control, and MRP) have been integrated into a single system. It also means that there is fallback from vendors, the production shop, and so on, when problems arise in implementing the production plan.

MRP II

Closed-loop MRP represents a significant achievement in terms of tying together the various separate functions of a production planning and control system. MRP II is, however, the final step in the evolution of MRP (at least, as it is currently conceived). This fourth step involves a link-up between the closed-loop MRP system and the financial systems of the company. Manufacturing resource planning is the name given to this combination. MRP II possesses the following two basic characteristics which go beyond closed-loop MRP:

(1) It is an operational and financial system.
(2) It is a simulator.

The operational and financial system makes MRP II a company-wide system, concerned with all facets of the business, including sales, production, engineering, inventories, and cash flows. In all cases, the operations of the individual departments are reduced to the same common denominator: financial data. This common base provides the company management with the information needed to manage it successfully. For example, raw materials on hand can be converted into their equivalent cost and summed over all stocks in inventory. Work-in-process can be evaluated by adding raw material costs to the cost of labor turned in against the particular part numbers and orders. Other operating data can be expressed in money terms by a similar calculation procedure.

MRP II is also a simulator which is intended to answer "what if" questions. It can be used to simulate the probable outcomes of alternative production plans and management decisions which are under consideration.

In essence, manufacturing resource planning is quite similar to the general model of a computer-integrated production management system (CIPMS). The CIPMS includes not only the operational system (MRP, inventory management, capacity planning) control module which is the connection with the company's accounting and financial systems.

QUESTIONS FOR REVIEW
(1) Describe the classes of MRP users.
(2) What are the basic characteristics of MRP II?

37 METHODS ANALYSIS [WORK STUDY]

INTRODUCTION

Knowledge of work study techniques alone is not sufficient for carrying out a successful work study program. The application of sound common sense is also important.

Methods analysis (or work study) has become a specialist function. It is now regarded as an essential management function. As the main objective of methods analysis is to improve the existing way of doing things by effecting change, regard should be paid to the reactions of all employees of any status who experience these effects.

Methods analysis should be integrated into the normal process of management and should not be left to specialists alone. It should not be started and applied haphazardly. It should be continuous. Managers should have sufficient knowledge of work study techniques, to ensure that they have proper executive control over their application and a full appreciation of their potentialities.

Every employee to a greater or less extent is interested in his own job. Unless work study techniques are properly applied, one should, therefore, expect to meet with considerable resistance at all levels in an organization. Firstly, there is the trade union to contend with. Many firms go to great trouble to make sure that at the outset of the application of work study, the trade unions are brought into the picture at the earliest possible moment. In consultations with the unions, there should be no hesitation in discussing honestly all the problems involved, and the possible repercussions and effects. Co-operation from the unions is of utmost importance to the successful application of work study techniques. Most unions now appreciate the immediate benefits which work study can give members by cutting out drudgery, frustration and unhealthy working conditions, by providing an opportunity for higher earnings, and indirectly by increasing the profitability of the company concerned and boosting the nation's economy as a whole. Naturally, there will be some fear that some employees will lose their jobs as a result of work or organizational re-organization. That is why it is all the more important that the fear of the employees should be alleviated through their union. Of course other means such as company newsletters and explanatory booklets.

On the other hand, managers may also feel threatened by the findings of the work study specialist, especially when well established routines are going to be upset. Therefore, it is necessary to involve the support of top management, without which any effort to initiate work study is fore-doomed to failure. If there were any danger of the existing management being made to feel incompetent or threatened by obvious improvements that can be made as a result of work study, some senior person should explain that it is a principle that there will be no recriminations or fault-finding as a result of the facts that have emerged from the study. It is also important that managers are made aware that methods analysis or work study is a tool of management, which should be accepted by all and not be looked upon with suspicion. It is of the greatest importance that lower management personnel, e.g., foremen, charge hands, and supervisors should be as closely involved with work study as possible. After all, these people are closest to the workers and can act as an excellent link between the methods analyst and the workers. They pass on to employees the details of what management requires to be done. They are responsible for implementing to a great extent the program of works on a monthly, weekly or daily basis, for output and quality of production and the proper utilisation of labor and raw materials, for safety standards for the methods used to carry out the work, and for the training of new workers and the retraining of existing workers.

All in all, everyone within the organization where methods analysis is undertaken should be convinced that:

(1) There is the economic necessity of reducing manufacturing costs.
(2) There are the advantages of systematic method study over occasional and haphazard attempts at method improvement.
(3) There are advantages in measuring work rather than relying on labor requirements sanctioned by custom and established possibly many years ago.
(4) All workers will get fair additional payment for additional work.
(5) Incentive schemes and the way in which the bonuses are worked out will be fair.

The methods analyst or work study practitioners should approach his job in a cautious, tactful manner. Every established organization has its set traditions and customs. By its very nature, methods analysis is a challenge to tradition and custom, and seeks to replace for these aspects of work a closer discipline relating to the facts. When the facts are proven they may be very different from what people believed was happening. There may be found to be obvious inefficiencies and wastage of time, material and effort, which were not at all obvious before the study was made. Therefore, it is important that whatever emerges in the way of pointers to possible improvement should not be used as a means of reproaching those who have been doing their best with insufficient information. These people should not be reproached for not having thought of the desirable changes. Instead, they should be given patient and clear exposition of why the changes are desirable.

Concern for the people involved is necessary if work improvements were to be really effective.

METHODS ANALYSIS
Methods analysis involves the development of more efficient work methods. More efficient work methods bring the firm closer to its goal of profit maximisation.

Therefore, in methods analysis, it is necessary to study the way in which something is being done, with a view toward developing a procedure which, if adopted, will serve to increase the firm's profits.

There are no hard and fast rules to methods analysis. However, it should be emphasized that the time to develop a good method is when preparations are being made to perform a task for the first time. Of course, the same principles will be involved as are involved in the analysis of an existing method. The only difference is that there is no need for describing the existing method because none exists. But every proposal for an original method should be analysed just as if it were an existing method.

Finally, it should be emphasized that though the human element is important to method improvement other aspects such as plant layout and materials handling should not be overlooked.

WORK MEASUREMENT - INTRODUCTION
A whole book can in fact be written about work measurement. However, we shall take an overview of work measurement. As mentioned earlier, work measurement is only one aspect of work study.

WHAT WORK MEASUREMENT IS
Work measurement is the procedure involved in measuring or forecasting the rate of output of an existing or newly designed operation, as well as in determining how much time is consumed for various productive and non-productive activities of a process, operation or job. It also involves the determination of standard times which represent the allowable time for the performance of work. Work measurement is a generic term and pertains to all techniques of time measurement of work systems. The results of work measurement are used by management as an analytical technique for the evaluation of work methods, determination of standard time values for given tasks, cost analysis and comparisons and the development of standard time data systems.

LABOR STANDARD
The actual work standard may vary considerably from the scientifically established industrial engineering standard. There is the impact of the informal organization, with its own communication network, system of authority, leaders, and work standards. Operations managers should not ignore the informal organization, and should attempt to influence the informal organization, communicate its work standards and at the same time attempt to influence the acceptance of formal standards by the informal work group.

A labor standard tells what is expected of an average worker performing under average job conditions. When establishing a labor standard, the following critical questions should be considered:

(a) How do we determine who is an "average" worker?

(b) What is the appropriate performance dimension to be measured?
(c) What scale of measurement should be used?

After answering these questions, we can use work measurement techniques to establish labor time standards.

People vary not only in such physical characteristics as height, arm span and strength, but in their working pace as well. To determine a labor standard, it is necessary to find an "average worker". But how do we find this person? If we choose one typical worker, he or she may not be typical in every respect. The best thing to do may be to observe several workers and estimate their average performance. It is necessary to trade off the costs of sampling and the costs of inaccurate standards. The total cost of establishing a standard is increased by the number of workers sampled and studied in depth. For example, if we study each of seven workers for one hour rather than each of three workers for an hour, the cost of studying performance (the sampling cost) more than doubles. The trade-off is that the more workers sampled and studied, the closer the performance standard should be to true "average" performance. There are also costs associated with inaccurate standards. Inaccurate standards can lead to inefficiencies, result in distorted product costs, and affect all the uses of standards. Of course an accurate standard can never be guaranteed, but if the number of workers studied is increased, the total costs of inaccuracy can be reduced. In trading off the costs of sample size and the costs of inaccuracy, a range of reasonably low total costs can be found.

There is yet another point about the concept of an average worker. Once average performance rates have been determined, the performance standard remains to be set. The question is whether the standard should be set at the average of total performances for the group, or at a level at which almost all the group can be expected to reach the standard. For example, should the standard be set at 22.25 units per hour, the mean performance, or at 14 units per hour, a figure that 95 % of the workers can be expected to reach? In fact, either is acceptable. Some engineers feel that adopting a minimum standard, that is, the second choice, encourages poor performances, and prefer to have about one-half the workers seeking but not attaining 100 % of the standard, that is, they suggest setting the standard at the mean performance (22.25 units per hour). Others feel that standards should be attainable by 90 % to 95 % of the workers. Both approaches can be used effectively.

HOW TO SET PERFORMANCE STANDARDS
When establishing work standards, management generally considers quantity to be the primary performance to be measured and quality the secondary standard. Quantity is usually measured as pieces per time period in manufacturing and service units per time period in service industries. For example, a lumber sawing operation may have standard performance set at 1,400 pieces sawed per hour and a bank teller may have standard performance measured and set at 25 customers served per hour. Quality

standards are often set as a percentage of defective units divided by total units and multiplied by 100. The sawing operation may have a quality standard of 1.0 % allowable defective units and the teller operation may allow a 0.05 % error in coin-counting. The cardinal points in determining dimensions of performance are:-

(a) The dimension must be specified before the standard is set.
(b) The standard and subsequent actual performance dimension must both be measurable.

Also a work measurement scale in which the normal performance is scaled at 100 % can be used. For example, if performance is 25 % above normal, the worker can be producing at 125 % of the normal scale.

ACCURACY
Obviously, experienced raters can set a standard more accurately than can inexperienced raters. Although even experienced raters make errors, the standards they set are generally found to have lower variability than standards set using only historical data. Raters should be used for work measurement, although setting a standard is not a finely developed scientific procedure and some errors are bound to occur.

TECHNIQUES OF WORK MEASUREMENT
There are six basic ways for establishing a time (work) standard, which are as follows:-

(1) Ignoring formal work measurement
(2) Using the historical data approach
(3) Using the direct time study approach
(4) Using the predetermined time study approach
(5) Using the work sampling approach
(6) Combining approaches (2), (3), (4) and (5)

Ignoring Formal Work Measurement
For many jobs in many organizations, especially in the labor-intensive service industries, formal labor standards are simply not set at all; there is no such thing as a fair day's work for a fair day's pay. The result is poor management or ineffective delegation. Workers may be blamed for poor performance and inefficiency, even though there is no explicit basis for criticism. If workers were not given specific, understandable goals, poor labor efficiency is bound to result. Usually because management has not established a work (time) standard, some informal standard is established by default. This informal standard, however, generally compares unfavourably with those set by work measurement techniques. Hence, formal work measurement should not be ignored.

Using The Historical Data Approach

This method assumes that past performance represents normal performance. In the absence of other formal techniques, some managers use past performance as their main guide in setting standards.

The advantages of using the historical data approach are that it is quick, simple, inexpensive, and probably better than ignoring the questions of establishing a work standard at all. The major disadvantage is that the past may not at all represent what an average worker can perform under average working conditions. Moreover, some of the historical data may reflect unusual working conditions or the performances of unusually capable or incapable workers. Unless management intuitively adjusts past performance data upward or downward before applying them as a standard, the historical approach may misrepresent average performance. However, in spite of these weaknesses, many organizations have used the method successfully to achieve goals of profitability, growth, and survival over extended periods of time.

Using The Direct Time Study Approach

Often called a time study, a stopwatch study, or "clocking the job", this technique is the most widely used method for establishing work standards in manufacturing. In this method, the industrial engineer simply studies a job, clipboard and stop-watch in hand. How does direct time study work? Basically, there are six steps in the procedure, which are as follows:-

(1) Observe the job being timed. This technique depends upon direct observation and is therefore limited to jobs that already exist. The job selected should be standardised, in terms of equipment and materials, and the operator should be representative of all operators.
(2) Select a job cycle. Identify the work elements that constitute a complete cycle. Decide how many cycles you want to time with a stop-watch.
(3) Time the job for all cycles. Workers behave in varying ways when their performances are being recorded, common reactions are resentment, nervousness, and slowing the work pace. To minimise these effects, repeated study, study across several workers, and standing by one worker while studying a job somewhere nearby, perhaps in another department, can be helpful.
(4) Complete the normal time based on the cycle times.
(5) Determine allowances for personal time, delays and fatigue.
(6) Set the performance standard (standard time) as the sum of observed normal time and determined allowances (the sums of steps (4) and (5)).

The following is another way to describe step (6) in the above procedure:

Standard Time = Normal Time - Allowance Fraction, where
Normal Time = (Average Cycle Time) x (Rating Factor)

Average Cycle Time = Time recorded for performing an element ÷ Number of cycles observed
Allowance fraction = Fraction of time for personal needs, unavoidable delays due to fatigue and:
0 < Allowance fraction < 1.0

Industrial engineers frequently use a rating factor when timing jobs. In essence, the engineer is judging the worker as 85 % normal, 90 % normal, or some other rating, depending on his perception of "normal". Needless to say, ratings of this kind depend on subjective judgements. For example, someone may ask why the sample was made up of five workers and fifteen cycles. This was because the engineer had intuitively judged that this sample was of sufficient size to give a reasonably accurate estimate of average time at reasonable cost; in direct time study, this is an accuracy/cost trade-off.

Using The Predetermined Time Study Approach

For jobs that are not currently being performed but are being planned, the predetermined time study approach is helpful in setting standards. Besides direct time study methods, predetermined time studies can also be applied to existing jobs. The bases of this technique are the stop-watch time study and time study from films. Historical data for tens of thousands of people making such basic motions as reaching, grasping, stepping, lifting, and standing are accumulated. These motions are broken down into elemental actual times, that are averaged by industrial engineers into predetermined standards, and published in table form. The procedure for setting a predetermined time standard is as follows:

(1) Observe the job or think it through if it is yet to be established. If you are observing the job, it is best to use a typical machine, representative materials, and an average worker performing the job correctly.
(2) Record each job element. Do not be concerned about elemental times, just thoroughly document all the motions performed by the worker.
(3) Obtain a table of predetermined times for various elements and record the motion units for the various elements. Motion units are expressed in some basic scale (a Sherblig scale is often used) that corresponds to time units.
(4) Add the total motion units for all elements.
(5) Estimate an allowance for personal time, delays, and fatigue in motion units.
(6) Add the performance motion units and allowance units for a standard job motion unit together and convert these motions units to actual time in minutes or hours. This total time is the resulting predetermined time standard.

Below is an example of a methods time measurement chart for the quantity standard. This chart shows the motions of the right and left hands, provides a code, and shows the TMU (Time Measurement Unit) motion units. It demonstrates how a predetermined time standard is set. The technique used here is called Methods Time Measurement (MTM) and is a widely accepted predetermined time study approach. The

MTM procedure allows one to observe the task, breaking it down into movements that are studied in depth and that have a predetermined average time. The MTM chart below is broken into several blocks for clarity. Toward the bottom of the chart there are the error allowance (placing an error card in the error box) and the allowance for personal needs (fatigue and unavoidable delay). The time allowance for personal needs, fatigue, and unavoidable delays is a standard industrial engineering allowance. A fifteen per cent allowance, which is a widely used allowance, is assumed for the task under consideration. If one TMU equals 0.00001 hour, the total MTM time per cycle, 397.9 TMU (as indicated in the chart below), is converted directly to 0.23874 minute per cycle. This is 4.188 units per minute, or 251 units per hour.

Right Hand	Code	TMU	Code	Left Hand
		14.2	R12D	Reach cards
		3.5	G1B	Grasp a card
		10.6	AP2	Apply pressure to separate
		3.5	T455	Turn card
		13.4	M12B	Move and focus eyes
Transfer card from other hand	G3	5.6		
Subtotal		50.8		Subtotal
Multiplied by 6		304.8		Multiplied by 6
		5.6	G3	Transfer cards from other hand
		13.4	M12B	Walk to final box
		4.0	D1E	Disengage cards
		15.0	WP(1)	Walk to start again
Subtotal		342.8		Subtotal
Error allowance		3.2		Error allowance
Subtotal		346.0		Subtotal
		51.9		15 % personal, fatigue, and delay allowance

METHODS TIME MEASUREMENT CHART FOR ESTABLISHING THE QUANTITY STANDARD

The principal advantage of predetermined time studies is that they eliminate non-representative worker reactions to direct time studies. Workers do not slow the pace or get nervous because the standard is set away from the work place in a logical, systematic manner. Since the workers are not anxious, disruptions on the shop floor are less severe with this technique than with direct time studies. The basic disadvantage of this technique is encountered early in its use. If some job elements are not recorded, or if they are recorded improperly, future timing will not be accurate. If job elements could not be properly identified and set forth in a table, they should be evaluated with the direct time study approach.

Using The Work Sampling Approach

Work sampling was pioneered in the 1930s, and is the most recently developed technique of the six being discussed here. In work sampling, stop-watch measurement is not carried out, whereas in many of the other techniques stop-watch measurement is resorted to. Instead, work sampling is based on simple random sampling techniques derived from statistical sampling theory. Its purpose is to estimate what proportion of a worker's time is devoted to work activities. Work sampling is carried out as follows:-

(1) Decide what conditions you want to define as "working" and what conditions you want to define as "not working." Not working consists of all activities not specifically defined as working.
(2) Observe the activity at selected intervals, recording whether a person is working or not.
(3) Calculate the proportion of the time a worker is engaged in work (P) with the following formula:

$P = x/n$ = Number of observations in which working occurred/Total number of observations

And standard times can be calculated with the formula below:

Standard time = Normal time/1 - Allowance fraction

The work sampling approach to job measurement is particularly suitable for service sector jobs, e.g., jobs in libraries, banking, health care, insurance companies, and government. A good deal of the accuracy of this technique depends on sample size. As is the case with any sampling procedure, there is a trade-off between larger sample size and increased accuracy, and the cost of increasing the sample size. It should be borne in mind that reliability and precision are important. A basic statistics book can assist one in setting the sample sizes for various reliability levels.

By including a concept known as rating or levelling performance, one can extend work sampling to include output standards. Once a job has been studied, the work study analyst has to decide whether the worker's performance was average, above average, or below average. If the analyst decides that performance was average no adjustment is made. If he decides that the worker was above average (measured as units/time period), the worker's rate is multiplied by a factor less than one. If the employee was found to be working below average, the rate is levelled to average by multiplying the observed performance by a factor of more than one. Levelling depends to a great extent on the industrial engineer's judgement and skill. There may be a lack of objectivity, and the result may be uneven from study to study. Moreover, the study has to be limited to a few workers. Furthermore, "willing" is a broad concept, and is not easily defined with precision. However, there are some obvious advantages with work sampling. It is simple, easily adapted to the service industries and the indirect laboratory output jobs, and an economical way to measure job performance. In fact, work sampling is a useful work measurement technique if it is used with discretion.

Combining Work Measurement Techniques

In practice, work measurement techniques are used in combination as cross-checks. One common practice is to observe a job, write down in detail all the job elements, and set a predetermined time standard. After this, one can check the history of performance on this job or similar jobs to verify that the predetermined standard is reasonable. To provide a further check, the job by elements and in total can be time studied. It should be remembered that no one work measurement technique is totally reliable. Setting a standard is not easy and requires great skill. There should be a cross-check whenever possible.

HOW ABOUT WORK MEASUREMENT FOR WHITE-COLLAR WORKERS?

Since white-collar jobs are typically labor-intensive and minimally automated, the same measurement techniques that are employed in the service industries will seem appropriate for them. The most suitable technique is probably a combination of historical data and work sampling. On the more routine white-collar jobs, predetermined time study can be a useful approach (when it can be used).

WORK MEASUREMENT USAGE

In 1976, a comprehensive survey on whether United States and Canadian industries used work measurement and, if they did, their reasons for having done so, had been carried out. Of the nearly 1,500 usable responses obtained in this survey, 89 % reported that they were using work measurement. This contrasts sharply with the 1956 survey, in which only 71 % of the 785 respondents reported that they had used work measurement to measure employee performance. The 1976 study also showed that work measurement had also been used for estimating and costing (89 %), establishing wage incentives (59 %) and production scheduling (55 %). This study makes it obvious that work measurement is by no means "dead" or "out-dated" in the industries. Apparently, work measurement is quite useful.

QUESTIONS FOR REVIEW

(1) What is work measurement?
(2) What is the relationship between work measurement and methods analysis? Which typically follows the other? Why?
(3) Explain how departmental and plant standards differ from individual job standards. Provide an example of each from an organization of your choice.
(4) Give two uses of time (labor) standards. Explain how the time standard could help a department within an organization of your choice for the two uses you have selected.

(5) Explain the predetermined time study approach to work measurement.

(6) Why would combinations of work measurement approaches be a good strategy in establishing a standard?

(7) Explain how you would proceed to set a standard for a group of seven draftsmen in a large architectural firm.

(8) Why are production/operations standards important?

HOW METHODS ANALYSIS SHOULD BE INTRODUCED

Two steps can be taken when introducing methods analysis. Firstly, we should make everyone aware of the management's intentions and create a favourable climate by educating members of management and selected workers and their representatives. The second step involves the actual use of work study techniques in various sections of the organization. Introducing work study is a lengthy process and something like a year may elapse between the first study and the first application. For work study to be successfully carried out, the support of the highest authority in the organization is necessary. In an industrial company this will usually be the board of directors. If such support is not available then the introduction of work study should be deferred until it is. Once support from the board of directors is assured, it is likely that there will be a senior official in charge in one locality, e.g., the works manager of the factory or the office manager of an insurance company, where it is proposed to introduce work study. It is necessary that this official becomes emotionally as well as intellectually convinced and determined on action. Lip service, which is unfortunately far too common, is not enough. He should have a sound appreciation of work study, and should attend a well-run management appreciation course.

The official should also meet others of like status and so realise the basic similarity of their mutual problems. No amount of reading and formal meetings can imbue the understanding necessary. There are a number of useful courses run by various organizations serving industry. Such courses may also be run internally if the organization in question is sufficiently large. These courses should be of at least one week's duration, though two weeks would be just nice. Great care should be taken to ensure that such courses are of high quality. Trouble, difficulties and a distorted picture of what work study can achieve will occur if the education of the official is inadequate. This applies to all members of middle management who can, directly or indirectly, influence its application. The application of sound common sense, a very difficult thing to teach indeed, is of great importance.

Middle Managers And Supervisors

The works or office manager should inform his middle managers and supervisors about the level of introduction of work study, giving them a broad outline and informing them that they will be receiving further information and training in due course. A lot depends on the manager's handling of this phase, as here he has to build up the confidence of his management team and imbue them with a spirit of enthusiasm for work study. He should especially emphasize that if there were changes and improvements none of these is going to be used as criticism of the existing or past activities of the team. This is vital to the morale and confidence of supervisors and middle managers.

UNION OFFICIALS AND MEMBERS

The intention on the management's part to introduce work study information should be discussed with the local trade unions as early as possible. There is everything to gain by doing so and a lot to lose by postponing this action. The manager or an appropriate company official should have an informal meeting with local union officials, with some of his own staff immediately concerned in attendance. It should be explicit that no commitment will be asked for a given, but that the meeting is a clear and practical demonstration that the manager intends to consult with the unions at all stages as further facts are established and progress is made. It should be made clear that dates are not yet known and staff will have to be appointed and trained and it may, therefore, be some months before management can make demonstrable progress.

The meeting should not be convened until management has a clear picture of the labor problems involved, and has prepared the framework of a policy which will guide their treatment of individual cases. A well-conceived policy should be flexible, bearing in mind that all eventualities cannot be foreseen, and should be far-sighted enough to avoid the danger of embarrassing precedents and "case law" due to inconsistent interpretation. One important point concerns redundancy, and policy on this should be very clear at the outset. It is important that the trade unions be informed about this and other aspects almost at the same time as the middle managers and supervisors, so as to prevent the almost inevitable distortions characteristic of the unofficial version of what is intended spreading throughout the organization.

First, all shop stewards and/or joint consultation representatives likely to be concerned should be given clear and practical help in order to gain understanding. This can be carried out by the manager at a meeting or meetings, or by providing facilities for the local trade union officials to tell their shop stewards. After this, notices should be posted throughout the organization to tell people of the management's intention to introduce work study. Middle managers and supervisors should be encouraged to discuss the subject with their work-people.

WHEN TO MAKE ANNOUNCEMENTS

Every possible effort should be expanded in building trust and confidence throughout the organization. This can only be possible when everyone is convinced that work study will bring benefits to all. The possible efforts of the ill-intentioned to sabotage the work study program should not be under-estimated. Careful timing and demonstrably honest explanation by the manager concerned in his various announcements can do much to prevent this from happening.

Middle managers, supervisors, local trade union officials, shop stewards and work people should all be informed on the same day and preferably in the same working period, say, the afternoon. For example, in the early afternoon, just after lunch, middle managers and supervisors can be officially informed at a meeting, followed by the local trade union officials at mid-afternoon and the shop stewards and/or joint consultation representatives in the later afternoon, with notices posted throughout the organization just before dismissal time.

The above example may represent a tight and difficult schedule but the resultant effect of everyone in the organization being informed officially is well worth the trouble. Once the management's intentions have been made known, efforts can be made to recruit or acquire the work study staff considered necessary for the implementation of the work study program. It is not advisable to recruit work study staff or to have them installed unless the management's intentions have been made known to everyone within the organization. All sorts of misunderstandings can arise among both management and their men if work study staff are installed without there being a clear statement of their function. Recruiting and training work study staff may take several months, during which time the other people within the organization, e.g., the local managers, the engineers, the foremen and supervisors, the charge hands, and the trade union officials and shop stewards, can be given training in work study.

It should be emphasized here that the whole methods analysis program should be carried out in such a way that it does not give any impression of being a "fault-finding mission"; everyone within the organization should be convinced of its necessity and the benefits it brings.

QUESTIONS FOR REVIEW
(1) What are the possible reactions toward a methods analysis program?
(2) What is the importance of methods analysis?
(3) How should methods analysis be introduced in an organization?

38 DEVELOPMENTS IN ELECTRONIC MANUFACTURING

Labor Problem

Manufacturing has always been a labor-intensive activity. Recruiting labor for electronic assembly work has been harder and harder because of the tight labor market. Moreover, certain jobs are so monotonous that labor turnover becomes high. Furthermore, the human worker has his quirks.

Mechanization And Automation

On the other hand, machines used in production or manufacturing activities have been playing a more and more important role. They not only perform a job more efficiently, they are super-workers when compared to humans.

The normal "conveyor line" system of electronic assembly has evolved to the automatic assembly lines or systems known as the "wave-soldering" systems, years ago. In wave-soldering, instead of having electronic components manually inserted and soldered onto the printed circuit boards by human hands, the inserting and soldering are done by the machine automatically which gives speed and accuracy resulting in comparatively little rejects. The normal active and passive components such as transistors, resistors and capacitors are still being used. The components are still soldered on only one side of the printed circuit board. Mass production is now carried out on a much larger scale, as compared to merely "hand-soldering" assembly.

The latest "in" technology in manufacturing is surface-mount assembly. This method of assembly depends on the "pick and place" robotic arm to do the job. An average surface-mount equipment could assemble 3,000 to 6,000 components an hour, with the "higher end", more expensive type capable of more than 10,000 components per hour. In this method of production, unlike the wave-soldering method, there is much more simplification and standardization so that a wide array of components could be automatically inserted and soldered onto the printed circuit boards. Many components are packaged in the same way (thus, it is not possible to tell at a glance what kind of components they are, unless we refer to their part numbers), which is "standardization" - this means that it is not possible to distinguish, e.g., between a diode and a resistor, without referring to their part numbers. Moreover, the components come in "tape and reel" form and are automatically fed into the surface-mount "pick and place" equipment like a reel of bullets being fed into the barrel of a machine-gun. This method of manufacturing necessitates that electronic components be specially manufactured and packaged, resulting in a new method of component manufacturing.

Moreover, the components (known as SMDs or surface-mount-devices) are now miniaturised (much smaller) and could be assembled on both sides of the printed circuit board (as compared to only one side for wave-soldering). This means that a much larger quantity of electronic components could be packed into a printed circuit board of the normal size. As consumer electronic equipment such as computers and calculators have "shrunk in size" for the convenience of the user (e.g., note-book and palm-top computers), their accompanying electronic circuitries have also to reduce in size. This is when surface-mount technology would play a useful role. This also necessitates the development of new types of soldering and desoldering tools for board-level rework or repairs, as the distance between components is now much reduced. Moreover, as the components are so much smaller, they are easily damaged when exposed to much heat (they over-heat faster), which make the selection of soldering and desoldering tools for rework/repair purposes critical as the heat from a soldering iron or desoldering iron tip could easily cause damage. This means that for some sensitive components such as integrated circuits, hot air is being used for soldering and desoldering work (in repairs or rework) instead of direct contact with a soldering or desoldering tool.

How does surface-mount work? The whole equipment comprises a robotic arm or arms, components feeders or trays and a drying oven. First, the printed circuit board is applied with a layer of paste on the side where the components are to be soldered, when it is in place and ready to receive the components. The robotic arm, which is programmable, would then pick the component from the feeder or tray and insert it on the printed circuit board, one by one. The component stays on the board firmly enough not to be dislodged as the paste keeps it in place.

After all the components or SMDs are in place, the whole board goes into an oven, which melts the paste so that the components become firmly entrenched on the printed circuit board. The method of heating in the oven is either by infra-red light or normal heat. However, of late, it has been found that the infra-red method of heating has a "light-shadow" effect problem, i.e., some components, being obstructed by the larger ones, would be out-of-reach of the infra-red light and thus would not be firmly in place.

Surface-mount is the newest in manufacturing technology, with the prices of the machine ranging from $200,000 to $750,000, or, more. And not all companies could afford it; certainly, it is out of the reach of many smaller companies. If for some reasons, your company wishes to go into surface-mount, you could consider the more established brands such as Dynapert, Panasonic and Fuji and the less established brands such as Technimount. There are of course many other brands. But, most companies would be concerned about the back-up technical support and the availability of spare parts provided by the supplier.

Computer-Aided-Manufacturing (CAM)
Now whole production lines could be automated and computerised, i.e., controlled and linked by computers. This is known as computer-aided-manufacturing (CAM).

Another relatively new concept in manufacturing is the flexible manufacturing system (FMS), whereby robots are well-utilized and the same production line could manufacture a variety of products. For example, a shoe manufacturing line could be converted into tooth-paste manufacturing.

With such new methods of manufacturing mass production could now proceed at an unprecedented pace, when production processes have become simplified further and the parts used have become more standardized.

This is also helped by the fact that many large circuits with many kinds of electronic components are now being "reduced to the size of the thumb" through advanced semi-conductor technology, which gives us many advanced ICs (integrated circuits). This obviously helps reduce the number of manufacturing processes (process simplification).

Modern Test Equipment

The normal method of testing assembled printed circuit boards is by getting a technician to use the multimeter, logic probe or oscilloscope.

However, with modern technology this may not be necessary. Modern in-circuit testers could accommodate the whole board, test the circuit by injecting voltages and signals, and give a computer print-out of the results, giving a mention of the defective sections as well.

Before the advent of computer-aided-design (CAD), an electronic designer has to bread-board assemble his just-designed circuit for testing. But CAD has made all this redundant. The designer could simulate and test the circuit he has designed with the computer only (using such software as HSPICE, i.e., he does not have to build a model to test based on his design. (What a saving of time!)

For circuits-on-a-chip (integrated circuits), universal programmers are available which allow one to program and test them. PALs, EPLDs, EEPLDs, GALs, DRAMs, SRAMs, microcontrollers, EPROMs and EEPROMs could be programmed and tested by a programmer about the size of a book. The "higher end" programmers could program up to about 32 ICs at one go very quickly and efficiently and could perform this job on up to about 3,000 ICs a day (say, eight hours) while the "lower end" ones could provide the same service for about 300 to 500 ICs a day (eight hours).

Moreover, many half-finished and fully-finished products have to be tested thermally as well as with humidity. The modern temperature test chamber is able to provide temperature cycles for "thermal shock" tests of parts and finished products, as well as for humidity tests. Such tests are important for equipment which could not afford to fail under varying weather conditions, e.g., in the case of automotive parts, wherein the car has to "bear" all kinds of weather condition - hot, cold, dry and wet weather conditions. The modern temperature test chamber is normally programmable and could be linked externally (optional) to a computer, through the RS 232 or GP1B interface card, for example. There are even special ovens for testing integrated circuits known as "burn-in" ovens.

Advances

The amount of equipment available for manufacturing is so amazing that the scope of this book would not be able to do enough justice to them.

As technology is still evolving, it is difficult to tell whether the equipment now used in manufacturing would be as useful in five to ten years' time.

The modern technical manager or engineering manager, whose main task is to support the production department, has to be keenly aware of the advances in manufacturing engineering.

39 MANAGEMENT OF MODERN TECHNOLOGIES

Today, the engineer armed with an engineering degree is just another student in the University of Life-Long Learning. He cannot afford to be complacent and rest on his laurels. The world is undergoing great technological advances and the scientist-engineer must be keen on keeping abreast of all the changes in technology.

The scientist-engineer has to really have a great interest in his field of specialisation if he wants to find out much about it and advances in it. Without this special, deep interest in the science, he could not expect to be able to keep up with the field. This is the most critical aspect of the scientist-engineer's make-up.

Secondly, he has to have access to scientist-engineers of similar interests so that they could test each other's ideas via discussions, seminars and research. They could even collaborate in research work, which would make the search for new knowledge a much more interesting one. "Two Heads Are Better Than One" is indeed here a truism.

Thirdly, these "hi-tech" personalities should be prepared to publish their findings or researches frequently and freely. The exchange of ideas through technical or scientific journals is an indispensable source of scientific inspiration as well as information. Today, such journals abound and are to be found in public libraries and libraries of academia.

Even scientists or engineers might have copies mailed directly to their homes or offices. This seems the fastest and easiest way of keeping abreast of advances in such a busy world, where we have to work long and hard hours and have precious little time to attend courses or seminars.

Scientists and engineers have to be altruistic and unselfish, for hoarding the knowledge would certainly slacken the pace of advance in knowledge. Moreover, technology has become so complex today that no one person could be an expert in everything. It would definitely require many experts to pool their skills and knowledge together to develop a technology. The one-man inventor is now a dying breed and is being replaced by a team of bright scientific researchers each of whom might be an expert in a field of their own.

It Takes More Than Humans To Develop A Technology
Even though the technology and the brains are there, a new technology could not be developed because the tools and equipment are not available as yet. For what is a new technology, if the tools for developing it are not available, but a pipe-dream? Charles Babbage, for example, was not able to complete his mechanical computer because the tools for manufacturing some of its parts were not available.

Moreover, the scientist-engineer is also bound by limits of the law and social-political factors which could be a serious hindrance to his work. If his work involves the use of ozone-depleting chemicals or "green-destroying" products, it is unlikely that it would receive much sanction or encouragement from the authorities or his fellow-beings. If he were involved in weapons of war or destruction, the pressure on his conscience and his efforts would be all the greater. Mankind, as sane beings, should however discourage, if not ban, the development of weapons of destruction.

The Good Of All Mankind
The scientist-engineer to be an effective one should keep the welfare and interest of fellow-beings the upper-most in his mind.

It would be a sheer waste of productive time to develop a thing that no one wants or finds useful which could only be regarded as a "crazy" invention, though it might be a very clever one. It is pointless to get a patent and be proud of it, if nobody buys the idea. You could make yourself prouder and probably happier by doing other things, for example, charitable work, which would surely be appreciated.

Inventions and advances in technology must be able to better the lives of our fellow-beings. In this respect, a cure for AIDS or a pill that stops the aging process would be a dream technology come true. For those who are in search of a perfect friend, how about a talking robot that could converse intelligently (even brilliantly) with you (as well as console you) on how to solve your personal problems?

Useful Knowledge
Advances in the field of electronics seem to affect our lives more than anything else. The laser discs, the camcorders, the computers, the karaoke sound systems and many more, are all there to titillate our tastes for the higher things in life. Advanced electronics technology, for example, could now give us Virtual Reality, wherein computers could give us an imaginary world which looks real, wherein an environment could be created by computerised electronic gizmos to make it look so realistic as to be believable.

The scientist or engineer working in the research laboratory conducting researches on electronic designs or doing chemical researches could now rely on Computer-Aided-Design or CAD nowadays to bring out new models and evaluate their performances (all via the computer and its dependent software) instead of having to spend time building an actual model which may be a time-consuming and futile effort (if the model does not work out). This certainly helps to hasten research and development work, making it possible for technology to advance at a much faster rate than ever before.

The effects of technological advances are cumulative. Advance in technology not only betters the quality of our lives, it also contributes directly towards its own advancement. To function well, the scientist-engineer should have a sound knowledge of computers and software such as Prolog, C, Excel, OSS and the like. It seems that such useful knowledge is rather indispensable for many research and development activities, where experiments or models based on new designs could be "computer-created" and evaluated instead of physically created, which could be a very time-consuming and frustrating effort.

As a matter of fact, the field of computer software development or computer programming is in itself a fast advancing field. Scientists-engineers should know how to use the latest software to enhance their own work, and learning how to use them well could be a life-long learning process.

Conclusion

The advances in technology today are so great as to put us all in awe of their effects on our lives. Automated telling systems, automated ticketing systems, local area networks (LANs), and many more, all bring smooth efficiency to the way our lives are being lived.

The modern technical manager must be keenly aware of new technologies and be ready to utilize them for the sake of development and improvement. Adopting a conservative attitude towards such things would make him a loser in the technological race.

An alert, keenly aware and open mind is today an indispensable item in the race for techno-supremacy. We hear about countries wanting to import technologies and scientists and engineers being sent abroad (especially to the U.S.) for training.

To expect some modicum of success in their fields of work, technical managers and engineers must not forget to enrol themselves as students in the University of Life-Long Learning.

40 THE HUMAN ASPECT OF TECHNOLOGY

The tools, the machinery, the computers, the robots, may play an important part in technology but the human element must never be overlooked.

The human element can not only be unproductive but can be "counter-productive" even; as we know, human beings can sabotage equipment, waste resources and carry out a long list of anti-social activities. Even, the slanderous words produced by the human vocal cords can blow great holes in the productivity effort.

No matter what, the human element can never be totally negated from the production process. You may go for full automation, full computerisation, but robots and computers can never move by themselves. If robots and computers can move by themselves, we will all be in danger. The human resource arrangement is still important in the fully automated factory environment.

Here, the author would like to talk about the humane management of human beings, a very important element. Union leaders here have in the past pleaded with managers (in the production process) to treat workers not as digits in the production process but as human beings. This is significant. The author feels that this has been in great part responsible for the large turnover of workers, both blue-collar and white-collar, in the country. This obviously upsets productivity. Employers complained about errant job-hopping amongst the staff. The government has taken corrective and penalising action against job-hoppers. But the problem is still there. Some job-hopping is forced or "coerced", some purely opportunistic.

From what the author can see, no amount of force can stop job-hoppers. If workers are not happy with the working environment, no amount of force or incentive can prevent them from leaving. If workers can be treated kindly and humanely, at least they will feel respected. Otherwise, all the time and training spent on them will be wasted when they leave. The ability to inspire workers and to draw the best from them is a quality that must be cultivated in all leaders of workers, line-leaders, foremen, supervisors, engineers and managers. Threat and fear are now out-of-place in the factory environment, and will never force workers to perform better if not forcing them to leave. Even money will often not induce them to stay.

The author quotes John Patton: "Direct incentives will increase production 20-50 per cent but there is an ingredient in the excellent companies in the potential that overshadows the productivity increase achievable through industrial engineering techniques. When we learn to manage people, the increased productivity will be likened to the relationship of a water wheel to nuclear energy."

Total productivity cannot be achieved unless workers are committed, motivated and trained to give of their best.

The best computer or the best robot will never perform at peak form if the programmer behind it is not committed to his job.

Productivity is a long term affair and management plays a very important part in the "productivity movement".

The following steps can be taken to enhance productivity in the factory:-

(a) Show Commitment
Verbal approval is insufficient. There must be manifestation that top management is concerned enough to do something.
(b) Provide Organizational Support
A top level steering committee can be appointed to plan and guide the overall productivity improvement effort.
(c) Plan For Productivity Movement
Productivity has to be built into the planning control and reward system.
(d) Provide Training
To improve productivity, it is necessary to develop a set of attitudes and skills. These can best be acquired by the utilization of training programmes.
(e) Publicize And Reward Productivity Improvement
Good results should be publicized as soon as they have been accomplished and good workers should be rewarded as promptly as possible.

Often, there is lack of effective communication between managers/supervisors and workers. Research has indicated that good labor-management relations has been a factor in higher productivity which was consistently regarded with priority by all employees. From the findings of a survey by the NCB and the Nomura Research Institute of Japan in October 1982, it was found that workers did seek to find fulfilment in their jobs and work-place policies and good working environment were cited by 34 per cent and 28 per cent of workers respectively as major reasons for being satisfied with their jobs.

This survey showed that there was no significant disparity in the responses of manager and workers. This means that the worker's job satisfaction regardless of his position in the organization was largely influenced by intrinsic rather than extrinsic factors associated with his job.

Almost half of the respondents (49.11 %) felt that human relations played an important part in job satisfaction. Managers supervisors and workers who were dissatisfied with the lack of care by management constituted 31.25 %, 29.17 % and 17.46 %. In other words, managers themselves were also dissatisfied with the way they were treated by their top management. The survey finding also confirmed that though awareness of the word "productivity" was high among the working population, workers generally tend to lack in-depth understanding of the concept of productivity. In fact, even managers and supervisors have not fully understood productivity.

The survey finding also indicated that managers and supervisors lack commitment to the movement. Their perception and degree of commitment regarding the movement was vague and superficial. Many did not fully realise the need for an overall productivity plan.

95 % of managers and supervisors in the survey stated that they would need to know more about how to improve productivity in their industries. Many also lacked in-depth knowledge pertaining to the various methods of productivity improvement. Workers were aware of their own limitations and the handicap posed by their lack of education and training. Many indicated their desire to upgrade themselves and were enthusiastic about learning new skills.

Thus, it can be seen and deduced from the findings of this survey that the human element is all important in manufacturing.

There is definitely a need for people-centered management. Workers should be given more opportunities in participating in decision-making and company affairs. They should be educated about productivity. Management must be able to inspire them to higher excellence and earn their trust and respect.

Unfortunately, most managements here tend to "cut corners" with workers. For example, when the government recommended National Wages Council increases for workers, only the civil service and unionised companies paid out these token increases to workers. Practically all non-unionised private companies ignored this recommendation and the unfortunate workers in these companies who have not the benefit of the protection of unions, did not enjoy a single cent of NWC increase. Workers, being human, would of course feel exploited.

There have been complaints by trade union leaders that workers have been merely treated as digits of production. They are hardly treated like proper human beings. Many managements look upon workers with nonchalance. Their attitude is "We pay you to do a job and we expect you to work hard". The fact that employers here always point the finger at job-hoppers is a manifestation of such an attitude: "We pay you to work. We spend time to train you. You should contribute something back to the company for all the time and effort we have spent on you".

But of course, as the survey findings have indicated, human relations can always be the cause of job-hopping. In other words, management had not set the environment right for their staff and unhappiness caused people to resign.

No management admitted their fault. Workers had no clout to blame them openly. Unions could only complain about management being impersonal and treating workers as mere units of production.

But management is all-powerful, and beyond questioning. They are only interested in profits. To them a worker, a human being, must be somebody who can help them to make money. If they cannot, they are no bloody good.

Yet come to think of it, a smile, a friendly pat, a warm gesture and some kind words can raise the workers' spirits to exalted heights and thus higher productivity. And this is not an expensive thing to do and can be done naturally and at practically no cost to management.

Workers should not be treated as extensions of the machines they are working with. A machine can be treated as a machine. But to treat a worker as a machine is not only not right but will never improve productivity and will even decrease productivity when workers' morale is undermined.

Even a machine will not function perfectly if not treated well; if you were to be rough and careless with the machine, if you do not practice "preventive maintenance", the machine will certainly break down more often.

The same too with human beings who work for you. It is much better to "prevent" human problems than to find "cures" for them. Prevention is always better than cure, no matter what.

41 CONCLUSION

Technology is a fast-moving field. The technical executive and his team have to keep abreast of the latest developments in technology. This makes their work challenging and refreshing. They have to look around and try to move forward all the time. They have to know what the latest technologies used by competitors are.

It pays for them to read widely the technical journals that are available. It also pays for them to attend courses and seminars to upgrade their know-how.

This book will somewhat enable the reader to improve on his technical, as well as managerial, know-how.

www.ingramcontent.com/pod-product-compliance
Lightning Source LLC
Chambersburg PA
CBHW081731170526
45167CB00009B/3785